# Natural Gas

# Natural Gas

## Fuel for the 21st Century

Vaclav Smil

**WILEY**

*Library of Congress Cataloging-in-Publication Data*

Smil, Vaclav.
    Natural gas : fuel for the 21st century / Vaclav Smil.
        pages   cm
    Includes bibliographical references and index.
    ISBN 978-1-119-01286-3 (pbk.)
1. Natural gas.   2. Gas as fuel.   I. Title.
    TP350.S476 2015
    665.7–dc23
                                                2015017048

A catalogue record for this book is available from the British Library.

ISBN: 9781119012863

Cover image: sbayram/iStockphoto

Set in 10.5/13pt Sabon by SPi Global, Pondicherry, India

1   2015

# About the Author

Vaclav Smil conducts interdisciplinary research in the fields of energy, environmental and population change, food production and nutrition, technical innovation, risk assessment and public policy. He has published 35 books and close to 500 papers on these topics. He is a Distinguished Professor Emeritus at the University of Manitoba, a Fellow of the Royal Society of Canada (Science Academy) and the Member of the Order of Canada, and in 2010 he was listed by Foreign Policy among the top 50 global thinkers.

## Previous works by author

China's Energy
Energy in the Developing World
  (edited with W. Knowland)
Energy Analysis in Agriculture
  (with P. Nachman and T. V. Long II)
Biomass Energies
The Bad Earth
Carbon Nitrogen Sulfur
Energy Food Environment
Energy in China's Modernization
General Energetics
China's Environmental Crisis
Global Ecology
Energy in World History
Cycles of Life
Energies
Feeding the World
Enriching the Earth
The Earth's Biosphere
Energy at the Crossroads
China's Past, China's Future

Creating the 20th Century
Transforming the 20th Century
Energy: A Beginner's Guide
Oil: A Beginner's Guide
Energy in Nature and Society
Global Catastrophes and Trends
Why America Is Not a New Rome
Energy Transitions
Energy Myths and Realities
Prime Movers of Globalization
Japan's Dietary Transition and Its Impacts
  (with K. Kobayashi)
Harvesting the Biosphere
Should We Eat Meat?
Made in the USA: The Rise and Retreat
  of American Manufacturing
Making the Modern World: Materials
  and Dematerialization
Power Density: A Key to Understanding
  Energy Sources and Uses

# Contents

# Preface

This book, my 36th, has an unusual origin. For decades, I have followed an unvarying pattern: as I am finishing a book, I had already chosen a new project from a few ideas that had been queuing in my mind, sometimes coming to the fore just in a matter of months and in two exceptional cases (books on creating and transforming the twentieth century) after a wait of nearly two decades. But in January 2014, as I was about to complete the first draft of my latest book (*Power Density: A Key to Understanding Energy Sources and Uses*), I was still undecided what to do next. Then I got an e-mail from Nick Schulz at ExxonMobil who is also a reader (and a reviewer) of my books, asking me if I had considered writing a book about natural gas akin to my two beginner's guides (to energy and to oil) published by Oneworld in Oxford in, respectively, 2006 and 2008.

I had written about natural gas in most of my energy books, but in January 2014, the idea of a book solely devoted to it was not even at the end of my mental book queue. But considering all the attention natural gas has been getting, it immediately seemed an obvious thing to do. And because there are so many components and perspectives to the natural gas story—ranging from the fuel as a key part of the United States' much publicized energy revolution to its strategic value in Russia's in its dealings with Europe and to its role in replacing coal in the quest for reduced greenhouse gas emissions—it was no less obvious that I will have to approach the task in my usual interdisciplinary fashion and that I will dwell not only with what we know but also describe and appraise many unknowns and uncertainties that will affect the fuel's importance in the twenty-first century.

I began to write this natural gas book on March 1, 2014, intent on replicating approach and coverage of the two beginner's guides: the intended readers being reasonably well educated (but not energy experts) and the coverage extending to all major relevant topics (be they geological, technical, economical, or environmental). But as the writing proceeded, I decided to depart from that course because I realized that some of the recent claims and controversies concerning natural gas require more detailed examinations. That is why the book is thoroughly referenced (the two guides had only short lists of suggested readings at the end), why it is significantly more quantitative and longer than the two guides, and why I dropped the word *primer* from its initial subtitle.

To many forward-looking energy experts, this may seem to be a strangely retrograde book. They would ask why dwell on the resources, extraction, and uses of a fossil fuel and why extol its advantages at a time when renewable fuels and decentralized electricity generation converting solar radiation and wind are poised to take over the global energy supply. That may be a fashionable narrative—but it is wrong, and there will be no rapid takeover by the new renewables. We are a fossil-fueled civilization, and we will continue to be one for decades to come as the pace of grand energy transition to new forms of energy is inherently slow. In 1990, the world derived 88% of its primary commercial energy (leaving aside noncommercial wood and crop residues burned mostly by rural families in low-income nations) from fossil fuels; in 2012, the rate was still almost 87%, with renewables supplying 8.6%, but most of that has been hydroelectricity and new renewables (wind, solar, geothermal, biofuels) provided just 1.9%; and in 2013, their share rose to nearly 2.2% (Smil, 2014; BP, 2014a).

Share of new renewables in the global commercial primary energy supply will keep on increasing, but a more consequential energy transition of the coming decades will be from coal and crude oil to natural gas. With this book, I hope to provide a solid background for appreciating its importance, its limits, and a multitude of its impacts. This goal dictated the book's broad coverage where findings from a number of disciplines (geochemistry, geology, chemistry, physics, environmental science, economics, history) and process descriptions from relevant engineering practices (hydrocarbon exploration, drilling, and production; gas processing; pipeline transportation; gas combustion in boilers and engines; gas liquefaction and shipping) are combined to provide a relatively thorough understanding of requirements, benefits, and challenges of natural gas ascendance.

# Acknowledgments

Thanks to Nick Schulz for starting the process (see the Preface).

Thanks to two Sarahs—Sarah Higginbotham and Sarah Keegan—at John Wiley for guiding the book to its publication.

Thanks again to the Seattle team—Wendy Quesinberry, Jinna Hagerty, Ian Saunders, Leah Bernstein, and Anu Horsman—for their meticulous effort with which they prepared images, gathered photographs, secured needed permissions, and created original illustrations, that are reproduced in this book.

# 1

# Valuable Resource with an Odd Name

Natural gas, one of three fossil fuels that energize modern economies, has an oddly indiscriminate name. Nature is, after all, full of gases, some present in enormous volumes, others only in trace quantities. Nitrogen (78.08%) and oxygen (20.94%) make up all but 1% of dry atmosphere's volume, the rest being constant amounts of rare gases (mainly argon, neon, and krypton altogether about 0.94%) and slowly rising levels of carbon dioxide ($CO_2$). The increase of this greenhouse gas has been caused by rising anthropogenic emissions from combustion of fossil fuels and land use changes (mainly tropical deforestation), and $CO_2$ concentrations have now surpassed 0.04% by volume, or 400 parts per million (ppm), about 40% higher than the preindustrial level (CDIAC, 2014).

In addition, the atmosphere contains variable concentrations of water vapor and trace gases originating from natural (abiogenic and biogenic) processes and from human activities. Their long list includes nitrogen oxides ($NO$, $NO_2$, $N_2O$) from combustion (be it of fossil fuels, fuel wood, or emissions from forest and grassland fires), lightning, and bacterial metabolism; sulfur oxides ($SO_2$ and $SO_3$) mainly from the combustion of coal and liquid hydrocarbons, nonferrous metallurgy, and also volcanic eruptions; hydrogen sulfide ($H_2S$) from anaerobic decomposition and from volcanoes; ammonia ($NH_3$) from livestock and from volatilization of organic and inorganic fertilizers; and dimethyl sulfide ($C_2H_6S$) from metabolism of marine algae.

*Natural Gas: Fuel for the 21st Century*, First Edition. Vaclav Smil.
© 2015 John Wiley & Sons, Ltd. Published 2015 by John Wiley & Sons, Ltd.

But the gas whose atmospheric presence constitutes the greatest departure from a steady-state composition that would result from the absence of life on the Earth is methane ($CH_4$), the simplest of all hydrocarbons, whose molecules are composed only of hydrogen and carbon atoms. Methane is produced during strictly anaerobic decomposition of organic matter by species of archaea, with *Methanobacter*, *Methanococcus*, *Methanosarcina*, and *Methanothermobacter* being the major methanogenic genera. Although the gas occupies a mere 0.000179% of the atmosphere by volume (1.79 ppm), that presence is 29 orders of magnitude higher than it would be on a lifeless Earth (Lovelock and Margulis, 1974). The second highest disequilibrium attributable to life on the Earth is 27 orders of magnitude for $NH_3$.

Methanogens residing in anaerobic environments (mainly in wetlands) have been releasing $CH_4$ for more than three billion years. As with other metabolic processes, their activity is temperature dependent, and this dependence (across microbial to ecosystem scales) is considerably higher than has been previously observed for either photosynthesis or respiration (Yvon-Durocher et al., 2014). Methanogenesis rises 57-fold as temperature increases from 0 to 30°C, and the increasing $CH_4:CO_2$ ratio may have important consequences for future positive feedbacks between global warming and changes in carbon cycle.

Free-living methanogens were eventually joined by archaea that are residing in the digestive tract (in enlarged hindgut compartments) of four arthropod orders, in millipedes, termites, cockroaches, and scarab beetles (Brune, 2010), with the tropical termites being the most common invertebrate $CH_4$ emitters. Although most vertebrates also emit $CH_4$ (it comes from intestinal anaerobic protozoa that harbor endosymbiotic methanogens), their contributions appear to have a bimodal distribution and are not determined by diet. Only a few animals are intermediate methane producers, while less than half of the studied taxa (including insectivorous bats and herbivorous pandas) produce almost no $CH_4$, while primates belong to the group of high emitters, as do elephants, horses, and crocodiles.

But by far the largest contribution comes from ruminant species, from cattle, sheep, and goats (Hackstein and van Alen, 2010). Soil-dwelling methanotrophs and atmospheric oxidation that produces $H_2O$ and $CO_2$ have been methane's major biospheric sinks, and in the absence of any anthropogenic emissions, atmospheric concentrations of $CH_4$ would have remained in a fairly stable disequilibrium. These emissions began millennia before we began to exploit natural gas as a fuel: atmospheric concentration of $CH_4$ began to rise first with the expansion of wet-field (rice) cropping in Asia (Ruddiman, 2005; Figure 1.1).

**Figure 1.1** Methanogens in rice fields (here in terraced plantings in China's Yunnan) are a large source of $CH_4$. Reproduced from http://upload.wikimedia.org/wikipedia/commons/7/70/Terrace_field_yunnan_china_denoised.jpg. © Wikipedia Commons.

Existence of inflammable gas emanating from wetlands and bubbling up from lake bottoms was known for centuries, and the phenomenon was noted by such famous eighteenth-century investigators of natural processes as Benjamin Franklin, Joseph Priestley, and Alessandro Volta. In 1777, after observing gas bubbles in Lago di Maggiore, Alessandro Volta published *Lettere sull' Aria inflammabile native delle Paludi,* a slim book about "native inflammable air of marshlands" (Volta, 1777). Two years later, Volta isolated methane, the simplest hydrocarbon molecule and the first in the series of compounds following the general formula of $C_nH_{2n+2}$. When in 1866 August Wilhelm von Hofmann proposed a systematic nomenclature of hydrocarbons, that series became known as alkanes (alkenes are $C_nH_{2n}$; alkines are $C_nH_{2n-2}$).

The second compound in the alkane series is ethane ($C_2H_6$), and the third one is propane ($C_3H_8$). The fossil fuel that became known as natural gas and that is present in different formations in the topmost layers of the Earth's crust is usually a mixture of these three simplest alkanes, with methane always dominant (sometimes more than 95% by weight) and only exceptionally with less than 75% of the total mass (Speight, 2007). $C_2H_6$ makes up mostly between 2 and 7% and $C_3H_8$ typically just 0.1–1.3%. Heavier homologs—mainly butane ($C_4H_{10}$) and pentane ($C_5H_{12}$)—are also often present. All $C_2$–$C_5$ compounds (and sometimes even traces of heavier homologs) are classed as natural gas liquids (NGL), while propane and butane are often combined and marketed (in pressurized containers) as liquid petroleum gases (LPG).

Most natural gases also contain small amounts of $CO_2$, $H_2S$, nitrogen, helium, and water vapor, but their composition becomes more uniform before they are sent from production sites to customers. In order to prevent condensation and corrosion in pipelines, gas processing plants remove all heavier alkanes: these compounds liquefy once they reach the surface and are marketed separately as NGL, mostly as valuable feedstocks for petrochemical industry, some also as portable fuels. Natural gas processing also removes $H_2S$, $CO_2$, and water vapor and (if they are present) $N_2$ and He (for details, see Chapter 3).

## 1.1   METHANE'S ADVANTAGES AND DRAWBACKS

No energy source is perfect when judged by multiple criteria that fully appraise its value and its impacts. For fuels, the list must include not only energy density, transportability, storability, and combustion efficiency but also convenience, cleanliness, and flexibility of use; contribution to the generation of greenhouse gases; and reliability and durability of supply. When compared to its three principal fuel alternatives—wood, coal, and liquids derived from crude oil—natural gas scores poorly only on the first criterion: at ambient pressure and temperature, its specific density, and hence its energy density, is obviously lower than that of solids or liquids. On all other criteria, natural gas scores no less than very good, and on most of them, it is excellent or superior.

**Specific density** of methane is $0.718 \, kg/m^3$ ($0.718 \, g/l$) at $0°C$ and $0.656 \, g/l$ at $25°C$ or about $55\%$ of air's density ($1.184 \, kg/m^3$ at $25°C$). Specific densities of common liquid fuels are almost exactly, $1,000$ times higher, with gasoline at $745 \, kg/m^3$ and diesel fuel at $840 \, kg/m^3$, while coal densities of bituminous coals range from $1,200$ to $1,400 \, kg/m^3$. Only when methane is liquefied (by lowering its temperature to $-162°C$) does its specific density reach the same order of magnitude as in liquid fuels ($428 \, kg/m^3$), and it is equal to specific density of many (particularly coniferous) wood species, including firs, cedars, spruces, and pines.

**Energy density** can refer to the lower heating value (LHV) or higher heating value (HHV); the former rate assumes that the latent heat of vaporization of water produced during the combustion is not recovered, and hence it is lower than HHV that accounts for the latent heat of water vaporization. Volumetric values for methane are $37.7\,MJ/m^3$ for HHV and $33.9\,MJ/m^3$ for LHV (10% difference), while the actual HHVs for natural gases range between $33.3\,MJ/m^3$ for the Dutch gas from Groningen to about $42\,MJ/m^3$ for the Algerian gas from Hasi R'Mel. Again, these values are three orders of magnitude lower than the volumetric energy density of liquid fuels: gasoline's HHV is $35\,GJ/m^3$ and diesel oil rates nearly $36.5\,GJ/m^3$. Liquefied natural gas ($50\,MJ/kg$ and $0.428\,kg/l$) has volumetric energy density of about $21.4\,GJ/m^3$ or roughly 600 times the value for typical natural gas containing $35–36\,MJ/m^3$.

Methane's low energy density is no obstacle to high-volume, low-cost, long-distance terrestrial **transport**. There is, of course, substantial initial capital cost of pipeline construction (including a requisite number of compression stations), and energy needed to power reciprocating engines, gas turbines, or electric motors is the main operating expenditure. But as long as the lines and the compressors are properly engineered, there is no practical limit to distances that can be spanned: multiple lines bring natural gas from supergiant fields of Western Siberia to Western Europe, more than 5000 km to the west. Main trunk of China's West–East pipeline from Khorgas (Xinjiang) to Guangzhou is over 4,800 km long, and eight major branches add up to the total length of 9,100 km (China.org, 2014). Moreover, pipelines transport gas at very low cost per unit of delivered energy and can do so on scales an order of magnitude higher than the transmission of electricity where technical consideration limit the maxima to 2–3 GW for single lines, while gas pipelines can have capacities of 10–25 GW (IGU, 2012).

Undersea pipelines are now a proven technical option in shallow waters: two parallel 1,224 km long lines of the Nord Stream project built between 2010 and 2012 between Russia and Germany (from Vyborg, just north of Sankt Petersburg to Lubmin near Greifswald in Mecklenburg-Vorpommern) to transport 55 Gm³/year were laid deliberately in the Baltic seabed in order to avoid crossing Ukraine or Belarus before reaching the EU (Nord Stream, 2014). Crossing deep seas is another matter: low energy density of natural gas precludes any possibility of shipborne exports at atmospheric pressure, and the only economic option for intercontinental shipments is to liquefy the gas and carry it in insulated containers on purpose-built tankers; this technique, still much more expensive than pipeline transportation, will be appraised in detail in Chapter 5. Methane's low energy density is also a disadvantage when using the fuel in road vehicles, and once again, the only way to make these uses economical is by compression or liquefaction of the gas (for details, see Chapter 7).

Low energy density would be a challenge if the only **storage** option would be as uncompressed gas in aboveground tanks: even a giant tank with 100 m in diameter and 100 m tall (containing about 785,000 m$^3$ or roughly 28 TJ) would store gas for heating only 500 homes during a typical midcontinental Canadian winter. Obviously, volumes of accessible stores must be many orders of magnitude higher, high enough to carry large midlatitude cities through long winters. The easiest, and the most common, choice is to store the fuel by injecting it into depleted natural gas reservoirs; other options are storage in aquifers (in porous, permeable rocks) and (on a much smaller scale but with almost perfect sealing) in salt caverns.

High **combustion efficiency** is the result of high temperatures achievable when burning the gas in large boilers and, better yet, in gas turbines. Gas turbines are now the single most efficient fuel convertors on the market and that high performance can be further boosted by combining them with steam turbines. When exiting a gas turbine, the exhaust has temperature of 480–600°C, and it can be used to vaporize water, and the resulting steam runs an attached steam turbine (Kehlhofer, Rukes, and Hannemann, 2009). Such combined cycle generation (CCG, or combined cycle power plants, CCPP) can achieve overall efficiency of about 60%, the rate unsurpassed by any other mode of fuel combustion (Figure 1.2). And modern natural gas-fired furnaces used to heat North America's houses leave almost no room for improvement as they convert 95–97% of incoming gas to heat that is forced by a fan through ducting and floor registers into rooms.

Little needs to be said about the **convenience of use**. The only chore an occupant has to do in houses heated by natural gas is to set a thermostat to desired levels (with programmable thermostats, this can be done accurately with specific day/night or weekday/weekend variations)—and make sure that the furnace is checked and cleaned once a year. Electronic ignition, now standard on furnaces as well as on cooking stoves, has eliminated wasteful pilot lights, and auto reignition makes the switching a one-step operation (turning a knob to desired intensity) instead (as is the case with standard electronic ignition) of turning a knob to on position (to open a gas valve), waiting a second for ignition, and then turning a knob to a preferred flame intensity.

Combination of these desirable attributes—safe and reliable delivery by pipelines from fields and voluminous storages, automatic dispensation of the fuel by electronically controlled furnaces, effortless control of temperature settings for furnaces and stoves, and low environmental impact—means that natural gas is an excellent source of energy for

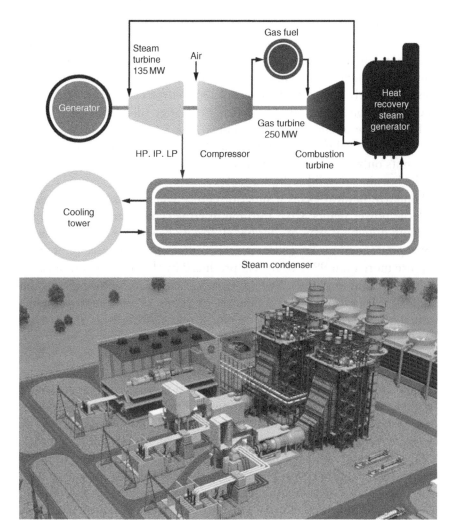

**Figure 1.2** Combined cycle gas turbine: energy flow and a model of GE installation. Reproduced courtesy of General Electric Company.

densely populated cities that will house most of the world's population in the twenty-first century. As Ausubel (2003, 2) put it, "the strongly preferred configuration for very dense spatial consumption of energy is a grid that can be fed and bled continuously at variable rates"—and besides electricity, natural gas is the only energy source that can be distributed by such a grid and used directly in that way.

As for the **cleanliness of use**, electricity is the only competitor at the point of final consumption. Combustion of pure methane, or a mixture of methane and ethane, produces only water and carbon dioxide $(CH_4 + O_2 \rightarrow H_2O + CO_2)$. There are no emissions of acidifying sulfur oxides (as already noted, $H_2S$ is stripped from natural gas before it is sent through pipelines), while heating houses with coal or fuel oil generates often fairly high emissions of $SO_2$. Moreover, coal combustion produces high concentrations of particulate matter (diameters of <10 µm, $PM_{10}$), and the smallest particles (diameter <2.5 µm, $PM_{2.5}$) are also fairly abundant when burning heavier liquids, while combustion of natural gas emits only a small fraction of the finest particulate matter compared to the burning of solid or liquid fuels.

Similarly, natural gas is a superior choice when generating electricity in large power plants. Coal burning in large central stations remains a globally dominant way of thermal electricity generation, and even with appropriate modern air pollution controls (electrostatic precipitators to capture more than 99% of fly ash produced by the combustion of finely pulverized coal in boilers; flue gas desulfurization to remove more than 80% of $SO_2$ produced by oxidation of coal's organic and inorganic sulfur), to generate a unit of electricity, it releases five to six times more $PM_{2.5}$ and $PM_{10}$ and in many cases more than 1,000 times as much $SO_2$ as does the combustion of natural gas (TNO, 2007).

No other form of energy has a higher **flexibility of use** than electricity: commercial flying is the only common final conversion that it cannot support, as it can be used for heating, lighting, cooking, and refrigeration; for supplying processing heat in many industries, powering all electronic gadgets and all stationary machinery; for propelling vehicles, trains, and ships; and for producing metals (electric arc furnaces and electrochemical processes). Natural gas shares the flying limitation with electricity—but otherwise, the fuel is remarkably flexible as its common uses range from household space heating and cooking to peak electricity generation and from powering compressors in nitrogen fertilizer plants to propelling LNG tankers (for details, see Chapter 4).

Because climate change and the future extent of global warming have become major concerns of public policy, **contribution to the generation of greenhouse gases** has emerged as a key criterion to assess desirability of fuels. On this score, natural gas remains unsurpassed as its combustion generates less $CO_2$ per unit of useful energy than does the burning of coal, liquid fuels, or common biofuels (wood, charcoal, crop residues). In terms of kg $CO_2$/GJ, the descending rates are approximately 110 for solid biofuels, 95 for coal, 77 for heavy fuel oil, 75 for diesel, 70 for gasoline, and 56 for natural gas (Climate Registry, 2013). Moreover, high-temperature coal combustion in large power stations also produces much higher volumes of the other two most important greenhouse gases: more than 10 times as much $CH_4$ and more than 20 times as much $N_2O$ per unit of generated electricity (details in Chapter 7).

**Reliability of supply** is perhaps best demonstrated by the fact that inhabitants of large northern cities hardly ever think about having their gas supply interrupted because such experiences are exceedingly rare. There may be a temporary problem with a distribution line bringing the gas to a house, and there are rare—and in a great majority of cases perfectly preventable—explosions. A widely reported explosion that leveled two apartment buildings, killed 7, and injured more than 60 in New York's East Harlem on March 12, 2014, is a good illustration of such an avoidable accident. Several residents said that they "smelled gas in the area for several days" (Slattery and Hutchinson, 2014). That unmistakable smell is butanethiol (butyl mercaptan, $C_4H_{10}S$) that is added to odorless natural gas in order to detect even tiny leaks by smell (humans can detect as little as 10 parts per billion of the skunk-like odor).

Nor are there any great uncertainties about the reliability of international gas supply. The most notable case of interrupted supply took place in January 2009 when Gazprom cut off all flows of Russian gas to Ukraine (also affecting the deliveries to more than half a dozen European countries whose gas must flow through the Ukrainian lines) due to the unpaid accumulated debt for previous deliveries (Daly, 2009). The flow was restored after 13 days, but by March 2014, new Ukrainian debts threatened a reprise of 2009, although Gazprom maintained that the EU consumers west of the Ukraine would not be affected. Another threat of export interruption (following Russia's annexation of Crimea and fighting in the Eastern Ukraine) was averted by a deal conclude in late October 2014. But this unique, albeit recurrent, threat is not a reflection of general natural gas trading practices, rather an exception due to an unsettled nature of Russia's relations with its neighbors in the post-Soviet era.

**Durability of supply** is, of course, the function of resources in place and of our technical and managerial capabilities to translate their substantial part into economically recoverable commodities—and natural gas ranks high on all of these accounts. Natural gas is present in abundance in the topmost crust in several formations, but until recently, only three kinds of gaseous hydrocarbons dominated commercial production. They are natural gases associated with crude oils, a very common occurrence in nearly all of the world's major oil reservoirs; these gases may be present in separate layers in a reservoir containing oil and gas, they may accumulate on top of oil as gas caps, or they may be (another common occurrence) dissolved in crude oil. Usual cutoff for associated gas is when a gas/oil ratio is less than $20,000\,ft^3$ per barrel (in SI units about $350\,m^3$ of gas per 100 l of oil).

In contrast, nonassociated natural gas comes from gas fields, that is, from hydrocarbon reservoirs whose gas/oil ratio exceeds the rate noted in the preceding paragraph. The two common categories of nonassociated natural gas are wet (also called rich) gas that is extracted from reservoirs where methane dominates but where heavier alkanes (NGL) account for a substantial share of hydrocarbon mixtures, while dry natural gas comes from reservoirs where prolonged heat processing produced gas mixtures containing more than 90% or even more than 95% of methane and only small amounts of ethane and other alkanes.

These three conventional resources continue to dominate global extraction of natural gas, but some countries and regions are now producing increasing quantities of gaseous fuel from coal beds and from shales, while extraction of gas from methane hydrates still awaits additional technical advances before becoming commercial. All resource totals are always only temporary best estimates, but in this instance, their magnitude guarantees generations of future use: the best recent assessments of recoverable resources indicate that the global peak of natural gas extraction is most likely no closer than around 2050 or perhaps even after 2070. Another reassuring perspective shows that during the past three decades (between 1982 and 2012), the world gas consumption rose 2.3-fold, but the global reserve/production ratio has remained fairly constant, fluctuating within a narrowband of 55–65 years and not signaling any imminent radical shifts.

This brief review shows that natural gas is an exceptional source of primary energy for all modern economies—not only because it combines a number of practical advantages and benefits but also because it either fits existing infrastructures or because any further expansion of extraction, transportation, and use can rely on proven and economical technical and managerial arrangements. This book will offer a concise, but systematic, review of key aspects that define and delimit the fuel's great potential during the twenty-first century: its biogenic origins, crustal concentrations, and widespread distribution; its exploration, extraction, processing, and pipeline transportation; its common uses for heat and electricity generation and as a key feedstock for many chemical syntheses; its pipeline and intercontinental exports and the emergence of global trade; diversification of its commercial sources; and the fuel's role in energy transitions and environmental consequences of its combustion.

Given the scope of the coverage and the length of the text, it is easy to appreciate that writing this book was a serial exercise in exclusion; at the same time and with every key topic, I have tried to go sufficiently below the shallows that now dominate the (web page-like) writings about technical, environmental, and economic complexities for nonexperts. That is why I hope that those readers who will persevere (and tolerate what I always think of as a necessary leavening by many requisite numbers) will be rewarded by acquiring a good basis for an unbiased and deeper understanding of a critical segment of modern energy use and of its increasing importance.

# 2

# Origins and Distribution of Fossil Gases

Many gases encountered in the biosphere are of organic (biogenic) origin: as already explained in the previous chapter, methane is actually the best example of a gas whose concentration is orders of magnitude higher than it would be on a lifeless Earth. Other gases generated by microbial metabolism include hydrogen ($H_2$), hydrogen sulfide ($H_2S$), carbon monoxide and dioxide (CO and $CO_2$), nitric oxide (NO), nitrous oxide ($N_2O$), and ammonia ($NH_3$), while photosynthesis releases huge volumes of oxygen ($O_2$), and all animal metabolism emits $CO_2$. But many of these gases are also of inorganic origins: CO from incomplete combustion of fuels and CO, $H_2S$, and $SO_2$ from volcanic eruptions— and methane was found in hydrothermal vents at the bottom of the Pacific Ocean and in other crustal fluids.

Given these realities, we have to ask if all fossil methane, or at least an overwhelming majority of its presence in the crust's topmost layer, is of biogenic origin (arising from transformation of organic matter) or if a significant share of it (if not most of it) is abiogenic, arising from processes that have not required metabolism of living organisms. These obvious questions have been asked and answered: debates about the origin of natural gas have been a part of a wider inquiry about the genesis of fossil hydrocarbons, with the consensus coming on the side of their organic provenance and with dissenters arguing that inorganic process is a better way to explain the genesis of vast gas volumes in the Earth's upper crust.

*Natural Gas: Fuel for the 21st Century*, First Edition. Vaclav Smil.
© 2015 John Wiley & Sons, Ltd. Published 2015 by John Wiley & Sons, Ltd.

I will recount this contest in the first section of this chapter before I get into a fairly systematic review of the known global distribution of natural gas resources and a few closer looks at major resource endowments in key producing countries and at some of the world's most notable natural gas fields. In the chapter's final section, I will assess the development of our understanding of the world's natural gas resources (whose total cannot be quantified with satisfactory accuracy) as well as the progression of natural gas reserves (whose appraisal, while far from perfect, offers a much more revealing understanding of realistic, economic prospects for exploiting the fuel).

## 2.1  BIOGENIC HYDROCARBONS

A widely accepted conclusion is that natural gas was formed by thermal decomposition (under temperatures of 150–200°C and higher), primarily of crude oil in reservoirs and, to a lesser degree, of organic matter in shales that were the source rock for the liquids (Hunt, 1995; Selley, 1997; Buryakovsky et al., 2005). But organic origins of oils are not as obvious as those of coals which contain fossilized trunks and imprints of leaves of trees that grew during the Carboniferous periods (360–286 million years ago) and in more recent pre-Quaternary eras (many lignites are younger than 10 million years). While coals are fossilized phytomass of the largest photosynthesizers, crude oils originated mostly from dead biomass of the smallest photosynthetic organism, single-cell phytoplankton composed mostly of cyanobacteria (photosynthetic bacteria that used to be known erroneously as blue-green algae) and diatoms (unicellular and colonial algae enclosed in siliceous cell walls), and from dead zooplankton (dominated by foraminifera, amoeboid protists with food-catching pseudopodia). In addition, there were secondary contributions of dead algal phytomass, invertebrate and fish zoomass, and assorted dead organic matter carried to oceans and lakes by rivers.

But there is no simple, direct path to transform microbial biomass into a pool of oil hydrocarbons: they originated mostly from nonhydrocarbon organic compounds (carbohydrates, proteins, and lipids) that were transformed by bacteriogenesis (microbial metabolism) followed by lengthy thermogenesis (heat processing within sediments). The process begins with gradual accumulation of dead biomass in coastal or lake sediments followed by aerobic microbial degradation releasing $CO_2$ and then by anaerobic fermentation releasing $CH_4$ and $H_2S$, and burial and compaction of organic matter in anoxic sediments produced complex mixtures of large organic molecules. These insoluble kerogens arise from different materials—mainly from algae in lacustrine settings, from plankton and some algae in marine formations, and mainly from higher plants in terrestrial environments—and they could be up to 10% (commonly just 1–2%) of the mass in shales and limestones, the usual source rocks of fossil hydrocarbons. Rate of their formation and their eventual concentration is an outcome of competing processes of accumulation, destruction, and dilution of organic matter.

Kerogens subjected to higher temperatures and pressures in buried sediments will be eventually degraded by a process that is very similar to deliberate actions in crude oil refineries: thermal cracking breaks up complex molecules, and kerogens are first transformed into dark (black or brown), near-solid bitumens. This transformative process, resulting in bitumen (dark, very viscous but inflammable organic matter) appears to be almost as old as life itself: residues of asphaltic pyrobitumen were found in Australia's Pilbara shales and were dated to 3.2 billion years ago (Rasmussen, 2000). Eventual cracking of bitumens produces lighter molecules of liquid hydrocarbons, and increasing temperatures mark the principal stages of the process (McCarthy et al., 2011).

Transformation of sediments rich in organic matter to sedimentary rocks (diagenesis) takes place usually at less than 50°C and at a depth of less than 1 km. Thermal cracking (catagenesis) is most effective between 65 and 150°C and typically at depths of 2–4 km. Heavier liquid molecules are formed at temperatures between 80 and 120°C, and catagenesis at higher temperature results in higher shares of lighter alkanes and in gradually increasing gas/oil ratio, and any prolonged thermal processing above 200°C produces only dry gas. Finally, metagenesis converts much of the remaining kerogen into methane and nonhydrocarbon gases at 150–200°C. Stolper et al. (2014), using an isotopic technique, delimited formation temperatures of thermogenic gases between 157 and 221°C (Figure 2.1).

Hydrocarbons in young (Cretaceous, 145–65 million years ago) reservoirs tend to be mostly heavy crude oils, lighter crude oils come from Jurassic or Triassic formations (younger than 250 million years), and the lightest alkanes are often of Permian or Carboniferous age (up to 350 million years old). As a result, the composition of hydrocarbon reservoirs spans a huge range from the extreme of highly viscous bitumen through heavy oils not accompanied by any gas to combinations of liquid and gaseous compounds (ranging from oils with only a small amount of dissolved gas to mixtures of oil, gas, and natural gas liquids) and to the other extreme of nearly pure methane with only a marginal presence of natural gas liquids.

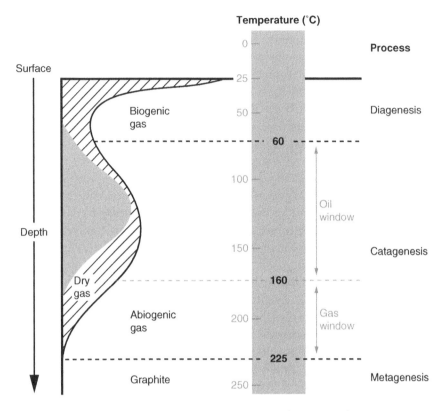

**Figure 2.1**   Diagenesis, catagenesis, and metagenesis.

Efficiency of this long process of transforming ancient biomass carbon into carbon in marketed fossil fuels can be quantified in terms of right orders of magnitude (Dukes, 2003). During coal formation, up to 15% of plant carbon ends up in peat, up to 90% of peat carbon is preserved in coal, and the now dominant opencast extraction can get up to 95% of coal in place from thick and level seams. This means that the overall carbon recovery factor (which is the fraction of carbon's original content in ancient phytomass that remains in extracted fuel) can be as high as 13%—or, restating this in reverse, that some eight units of ancient carbon (with the most common range of 5–20 units) was transformed into one unit of carbon in marketed coal.

In contrast, carbon preservation rates were much lower in marine and lacustrine sediments, and hydrocarbon recovery rate is only rarely close to 50%. As a result, the overall recovery factor for carbon sequestered in crude oil carbon could be as high as nearly 1% and lower than 0.0001%. Common rate of 0.01% means that 10,000 units of ancient carbon in aquatic biomass were transformed into 1 unit of carbon in marketed crude oil, and subsequent catagenesis and metagenesis, even if operating with 80% efficiency, would result in more than 12,000 units of carbon in ancient aquatic phytomass and zoomass to produce a unit of carbon in methane and natural gas liquids.

But what if hydrocarbons were of inorganic, rather than biogenic, origin? That was assumed by Dmitri Ivanovich Mendeleev, Russia's leading nineteenth-century chemist, and that has been an alternative to the biogenic explanation offered by the so-called Russian–Ukrainian hypothesis about the abiogenic formation of oil and gas in abyssal environments. The theory was first formulated during the early 1950s, and it had been championed by its proponents in the USSR and by some Western geologists (Kudryavtsev, 1959; Simakov, 1986; Glasby, 2006). According to the theory, formation of highly reduced hydrocarbons with high energy content from highly oxidized organic molecules with low energy content would violate the second law of thermodynamics, and high pressures deeper in the Earth's mantle are the best explanation for the formation of such reduced molecules.

Porfir'yev (1959, 1974) had also argued that abiogenic formation of giant oil fields is a better explanation of their origins than assuming truly gigantic accumulations of organic material that would be needed to create such structures. Modern interpretation of the abiogenic theory was summarized in a paper published in the *Proceedings of the National Academy of Sciences* and authored by Jason F. Kenney (a leading American advocate of inorganic origins of hydrocarbons) and his Russian colleagues. They recounted the experiments that produced a range of petroleum fluids in an apparatus replicating pressure (50 MPa) and temperature (up to 1500°C) 100 km below the surface (Kenney et al., 2002).

A kindred alternative was advocated by Thomas Gold, an American astrophysicist (Gold, 1985, 1993). Gold noted the presence of abiogenic hydrocarbons on planetary bodies devoid of life and maintained that methane can form by combining hydrogen and carbon under high temperatures and pressures in the outer mantle, and after this mantle-derived methane migrates it is then converted to heavier hydrocarbons in the upper layers of the Earth's crust. If true, this would have two profound consequences: hydrocarbons created by degassing from the mantle would be much more plentiful than is indicated by estimates attributed to biogenic formation; and perhaps the most intriguing consequence of abiogenic hydrocarbon formation, existing reservoirs could be gradually recharged (albeit at a very slow rate) by continuing formation of oil and gas (Mahfoud and Beck, 1995; Gurney, 1997).

But the alternative explanations of oil and gas origins have not aged well. Once the Russian–Ukrainian theory became better known abroad (starting in the mid-1970s), it was repeatedly dismissed by most of the European and North American petroleum geologists who favor the consensus explanation that excludes any major contribution by abiogenic origins of hydrocarbons. This consensus view is strongly supported by geological and geochemical evidence, and it has been strengthened by the use of the latest analytical methods (Glasby, 2006; Sephton and Hazen, 2013). I will note here half a dozen of major realities that undermine the abiogenic hypothesis.

The upper mantle is too oxidizing to allow the persistence of significant amounts of $H_2$ or $CH_4$; hydrocarbons formed from outgassed methane should be (but are not) concentrated largely along major tectonic discontinuities (faults and convergent zones); modern understanding of fluid and gas migration explains many previously puzzling hydrocarbon occurrences; biomarkers (including porphyrins and lipids) are derived from organic molecules; and isotopic analyses show a match between carbon isotope ratios in hydrocarbons and in terrestrial and marine plants. Lollar et al. (2002) used isotopic analysis of carbon and hydrogen to show a clear distinction between thermogenic and abiogenic hydrocarbons, and due to the absence of appropriate isotopic signatures in economically important reservoirs, they ruled out the occurrence of globally significant abiogenic alkanes.

This does not mean that there are no hydrocarbons of inorganic origin and that we have satisfactory explanation for the formation of all major hydrocarbon deposits. Gold may have overstated his case for abiogenic methane, but he was right when he posited the existence of what he called deep hot biosphere (Gold, 1998), assemblages of extremophilic bacteria living deep underground, up to several km below the Earth's surface, others deep below the deep-sea bottom. That claim was initially dismissed by the prevailing scientific consensus only to be confirmed later by incontrovertible and amazing findings of such organisms (Reith, 2011). And Milkov's recent explanation of the formation of giant gas pools in Western Siberia is an excellent illustration of complexities involved in the genesis of hydrocarbons (Milkov, 2010).

Dry gas pools in the northern part of the West Siberian Basin contain about 11% of the world's conventional gas reserves and account for 17% of current global gas extraction, but none of the proposed hypotheses of their origin (thermogenic gas from deep source rocks, microbial gas from dispersed organic matter, thermogenic gas from coal) are consistent with actual molecular and isotopic composition of extracted gases. Milkov argues that a significant (but unquantified) share of those shallow dry gases is the result of methanogenic biodegradation of petroleum rather than the outcome of thermogenesis. This also illustrates the complexities of postformation hydrocarbon histories. Oil and gas originating in kerogen-rich source rocks will almost invariably migrate through porous and hence permeable reservoir rocks and will be held in place by impermeable traps to hold the liquid in place: we find gas (and oil) only in those places where all of these conditions are appropriately combined.

## 2.2   WHERE TO FIND NATURAL GAS

Systematic exploration and assessment of fossil hydrocarbons present in the Earth's uppermost crust resulted in the fundamental division into conventional and nonconventional natural gas resources, that is, the ones that are relatively easy to extract and those that are much more difficult and hence much more costly to recover or that cannot be tapped at all with existing production techniques. The first category of methane-dominated gas mixtures is made up of two distinct resources. Historically, the most common source of gaseous hydrocarbons has been associated gas, that is, a gas dissolved in crude oil and sometimes also forming caps in oil reservoirs.

The world's largest oil field, Saudi al-Ghawār, is an excellent example of a reservoir containing both liquid and gaseous hydrocarbons: besides producing annually about 250 Mt (10.5 EJ) of oil, it also yields about 21 Gm$^3$ (750 PJ) of associated natural gas (Sorkhabi, 2010). And in 1971, 30 years after the reservoir began producing crude oil, a large pool of nonassociated gas was discovered below the oil-bearing layers at a depth of 3–4.3 km, and this deep reservoir now produces annually about 40 Gm$^3$ (1.4 EJ) of nonassociated gas. And other Middle Eastern oil supergiants are even richer in natural gas.

Al-Ghawār's oil-to-gas reserve ratio is greater than 30; for al-Burqān, the world's second largest hydrocarbon reservoir, it is only about 5; for the Iranian Aghajari, it is about 3; and for Marun (Iran's second largest oil field), it is less than 1.5. And in the Williston Basin of North Dakota, extracting oil from Bakken oil shale by horizontal drilling and fracturing has been accompanied by so much associated gas that, given the rapid growth of production and lack of adequate transport capacity, gas flaring has been so extensive that nighttime satellite images show large patches of light rivaling such large metropolitan areas as Minneapolis or

Denver (see Chapter 7). In other basins, large stores of gas and oil may be found separately: Algerian Hassi R'Mel (discovered in 1958) is a part of a large Saharan Triassic basin that also contains supergiant oil field at Hassi Messaoud.

As the post-WW II demand for natural gas increased and as new large-diameter pipelines enabled long-distance transport and exports, the share of natural gas originating from oil fields began to decline, and currently, most of the world's natural gas extraction comes from gas reservoirs. Some of them contain almost pure methane, with the world's largest accumulation of natural gas in a large group of giant fields in the West Siberian Basin being the best example of this high purity (Figure 2.2). These Cenomanian fields (dating to oldest Late Cretaceous epoch of 100.5–93.9 million years ago) contain gas averaging 97.95% $CH_4$, 0.23% $C_2H_6$, 1.58% $N_2$, and 0.24% $CO_2$ with traces of He and Ar, a mere 0.019% $H_2$, and no $H_2S$ (Milkov, 2010). This composition makes the gas virtually pipeline-ready without any need for processing. But a much more common occurrence is methane with variable but relatively high shares of natural gas liquids (ethane to pentane): the former category, where more than 85% (and as much 95%+) of the volume is $CH_4$, is commonly known as dry gas; when $CH_4$ falls below 85% and the mixture is rich in natural gas liquids, it is wet gas.

Nonconventional gas is present in four major formations. Enormous quantities of gas are locked in tiny bubbles in shales, a resource whose recent commercial exploitation in the United States has been so extensive and so rapid that many commentators have called it a revolution. Natural gas is also locked in tight formations in exceptionally hard, impermeable rocks. Presence of methane in coal beds has been an unwelcome reality due to an increased risk faced by miners as a result of sudden underground explosions—but in many areas, that gas offers a much cleaner alternative to highly polluting coal. And natural gas is also concentrated in massive underground and undersea deposits of methane hydrates (or clathrates), with the gas trapped inside a cage-like lattice of ice.

As yet there is no commercial recovery of methane hydrates, and most gas-producing countries tap only one or two of the remaining gaseous resources. The United States extracts all of them, and output statistics show the relative importance of these four principal sources: in 2011, 43% of gross gas withdrawals came from gas wells, 21% from oil wells, 6% from coal bed wells, and 30% from shale gas wells and tight gas deposits, the category whose output was lower than coal bed gas as recently as 2007 (USEIA [US Energy Information Administration], 2014a). I will take a closer look at all nonconventional sources of natural

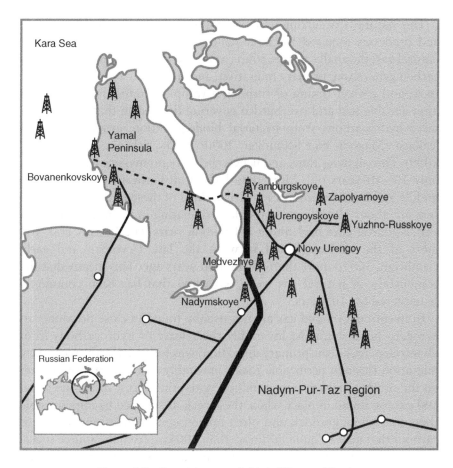

**Figure 2.2**    Supergiant gas fields in Western Siberia.

gas in Chapter 6, and in this section, I will explain where we search for conventional gas.

Finding fossil hydrocarbons became easier once we came to understand their biogenic origins beginning with large accumulations of kerogens. Looking for hydrocarbons means focusing on sedimentary basins where layers rich in organic matter became source rocks, where permeable, porous sediments became reservoir rocks and where, due to structural or stratigraphic reasons, these porous rocks became sealed at the right time by impermeable cap rocks. Magoon and Dow (1994), building on at least half a dozen previous syntheses, related all of these elements and processes in terms of the petroleum system, later renamed the total petroleum system (TPS).

This naturally occurring hydrocarbon system entails all the elements and processes required for large-scale accumulation of crude oil and natural gas: thermally mature source rock; a complex sequence of hydrocarbon generation, primary migration, and storage in a porous reservoir rock; and correct timing of forming a trapping configuration with an impermeable seal and overburden covering the system that help to preserve hydrocarbons from bacterial biodegradation, evaporation, and leakage (Magoon and Schmoker, 2000; Peters, Schenk, and Wygrala, 2009). This all takes time, and hence, there are no massive gas reservoirs only $10^3$–$10^5$ years old. More than half of all kerogens come from the middle and late Mesozoic era: nearly a third of these oil and gas precursors are about 100 million years old (originating in the mid-Cretaceous era), a quarter goes back about 150 million years (late Jurassic era), and most of the rest originated between the late Devonian and early Cambrian, 360 to more than 550 million years ago. This means that the combustion of natural gas oxidizes carbon that has been commonly sequestered for $10^8$ years.

Conventional oil and gas are not usually found in close proximity to kerogens because almost invariably the release of hydrocarbons from their source rocks (the primary migration) was followed by the secondary migration through permeable rocks, inevitably a very slow progression on the order a few km over a million years; the largest reservoirs of oil and gas are found in place where these rock are capped by impermeable traps that allow gradual, and often immense, accumulations of hydrocarbons that may be quite far (more than 200 km) from the source rocks. In an overwhelming majority of cases, oil and gas reservoirs (much like water reservoirs) are not large caverns filled with hydrocarbons; they are rocks whose porosity and permeability allowed first the ingress and then a lasting storage of large volumes of liquids and gases that could eventually, after drilling, flow upward through wellbores.

As a result, hydrocarbon reservoirs range from large, thick, and contiguous formations to small, shallow, and broken-up storages. More than half of all reservoir rocks are of Mesozoic age, almost two-fifths are Cenozoic, and a small remainder are the oldest Paleozoic sediments, with carbonates and sandstones dominant (Slatt, 2006). Carbonate rocks harbor some of the world's most notable hydrocarbon reservoirs. They were deposited in shallow seas either through the precipitation of calcium and carbonate ions or the biomineralization by marine organisms (mainly by coccolithophorids and foraminifera). Limestones are composed of calcite and aragonite, both being $CaCO_3$, calcium carbonate, only with different crystal structure. Dolostones are composed of more porous dolomite, $CaMg(CO_3)_2$. As for their age, many of the world's largest hydrocarbon reservoirs, including the enormous accumulation in the Persian Gulf, Arabian Peninsula, and West Texas, are in Cretaceous, Jurassic, or Permian carbonates deposited 66–286 million years ago.

Clastic sediments (including sandstones, siltstones, and shales) arise after transport and redisposition of fragmented rocks. This can take place in meandering rivers and in braided streams, with Triassic age sandstones of the Prudhoe Bay on the northern coast of Alaska being an excellent example of the latter process. In China, there are many sediments of lacustrine origin, deposited from ancient large lakes in Qaidam and Junggar Basins in China's western interior, and the country's largest gas field, Sulige in Ordos (discovered in 2000), contains almost $1.7\,Tm^3$ of tight gas (of which nearly $900\,Gm^3$ are recoverable) in Carboniferous and Permian deltaic sandstones and in Upper Permian lacustrine mudstones (Yang et al., 2008). Reservoirs in deltaic deposits and submarine fans include fields along the northern shores of the Gulf of Mexico, in the Niger's delta in the Gulf of Guinea, Trinidadian fields in the Orinoco's delta, and the fields in the southern part of the Caspian Sea.

High porosity of reservoir rocks can be the result of initial rock formation or subsequent fracturing and reformation, and effective porosity (total volume of interconnected pores) is usually much higher in sedimentary formations (primary porosity) than in highly fractured igneous or metamorphic rocks (secondary porosity). Porosity of sedimentary rocks (pore volume/total volume of reservoir rock) is commonly between 10 and 20%; in some smaller volumes, it can be higher than 50%; and 8% is about the lowest level allowing conventional hydrocarbon extraction. In the United States, it ranges between 1 and 35% in carbonate reservoirs and averages 12% in limestones and 10% in dolomites (Schmoker, Krystinik, and Halley, 1985). Examination of nearly 37,000 producing reservoirs shows expected correlation with depth and age: older and deeper reservoirs are less porous, with the youngest siliciclastic reservoirs averaging more than 20% and pre-Cambrian carbonate reservoirs staying mostly below 10% (Ehrenberg, Nadeau, and Steen, 2009).

Permeability, the capacity to transmit fluids and gases through porous materials, is the other key attribute, and it is primarily determined by the size of grains and by the properties of the medium. Its effective rate is usually measured in darcy units (D, after Henry Darcy). Impermeable rocks that form reservoir caps (NaCl, $CaSO_4$) have permeability of just $10^{-6}$ to $10^{-8}\,D$. In contrast, highly permeable reservoir rocks have permeabilities in excess of 1,000 mD (1 D), typical permeabilities of hydro-carbon reservoirs are 10–100 mD, but in some reservoirs, permeability may differ by orders of magnitude between different layers and areas.

Traps that contain gases and liquids in place are classified either as structural (arising, often on a vast scale, through deformation of the crust) or stratigraphic (usually smaller, formed either by gradual accumulation of impermeable rocks or by sudden shifts) and must be topped by impermeable layers. Anticlines, large convex (arched, domed) folds, are by far the most important structural traps confining roughly 80% of the world's largest hydrocarbon reservoirs (Figure 2.3). Among the best traps are anticlines created by rising salt domes and, in addition, they also appear often together with impermeable anhydrite ($CaSO_4$) or gypsum ($CaSO_4 \cdot 2H_2O$), making perfect seals.

The largest anticlines arise by compression along the convergent zones of tectonic plates: the Zagros fold belt of Iran is their foremost example. Among the supergiant gas fields, sandstone anticlines include the West Siberian Urengoy, Kuwaiti al-Burqān, Indonesian Minas on Sumatra, Shatlyk in Turkmenistan, Alaska's Prudhoe Bay, as well as several layers of the Qatari/Iranian North Dome/South Pars field. Among the major North Sea fields, those producing both oil and gas include British Brent (discovered in 1971) and Norwegian Ekofisk (1969), Statfjord (1974) and, above all, Troll, a sandstone anticline discovered in 1979 at a depth of 1,400 m and holding nearly 130 Gm³ of recoverable gas.

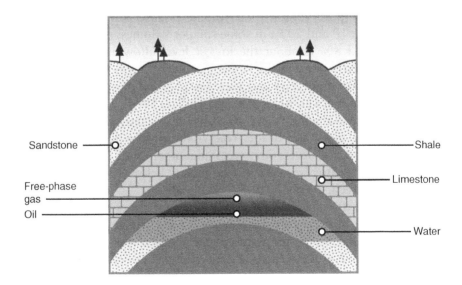

Figure 2.3   Anticlines.

Putting all of these attributes together, the standardized descriptions of four of the world's largest natural gas fields read as follows. The world's largest accumulation of natural gas, the North Dome/South Pars in the Persian Gulf between Qatar and Iran (discovered in 1971), is a highly stratified Permo-Triassic carbonate reservoir with anticlinal trap about 2,900 m below the sea bottom, 200 m thick with porosity of 9.5% and permeability of 300 mD whose recoverable reserves are at least 22–24 Tm$^3$; deep Silurian (Qusaiba) shale is the most likely source rock, with gas coming from two zones (Upper Dalan and Kangan), about 200 m thick and 2.9 km below the sea bottom, with porosity of about 10% and permeability of 300 mD (Esrafili-Dizaji et al., 2013). The gas is abnormally pressured but it contains high levels of $H_2S$ as well as of $CO_2$.

The world's second largest natural gas field, Galkynysh in southern Turkmenistan in the Amu Darya Basin (discovered in 2006), is actually a cluster of fields (South Yolotan, Osman, Minara, and Yashlar) with anticlinal traps. The former number two, Urengoy in the Western Siberian Basin (discovered in 1966), is a Cretaceous sandstone reservoir with anticlinal trap, about 70 m thick and 1–3.1 km below the ground, with porosity between 14 and 20% and permeability of 7–170 mD (Milkov, 2010; Li, 2011). Its source rock is the late Jurassic Bazhenov shale formation (which also contains tight gas), and as in virtually all reservoirs in the basin, the gas is nearly pure methane with no $H_2S$.

Europe's largest continental field, Groningen near Slochteren in northern Netherlands (discovered in 1959), is a Permian sandstone reservoir with uplifted and faulted trap, about 160 m thick and 2.7–3.6 km below the ground, with porosity of 15–20% and with permeability ranging over four orders of magnitude from less than 1 mD to as much as 3000 mD (Li, 2011). Its source rocks are about 300 million years old carbonaceous shales. America's largest natural gas field, Hugoton–Panhandle in Kansas, Texas, and Oklahoma (discovered in 1918), is a sprawling (nearly 24,600 km$^2$) formation whose reservoir rocks include carbonates, dolomites, and limestones, with a complex, giant stratigraphic trap. Its most prominent source rock is late Devonian shale (older than 350 million years), the reservoir thickness is about 170 m, the average porosity is 9.2% (but modal value is just 7%), and the permeability has a huge range from 0.0001 to 2,690 mD; the reservoir has subnormal pressure and contains gas of variable composition (Sorenson, 2005; Dubois et al., 2006).

When the USGS assessed the world's oil and gas resources (Magoon and Schmoker, 2000), the five largest TPS ranked by their known gas volume were the Northern West Siberian Mesozoic Composite TPS (including Urengoy, Yamburg, and Bovanenkovo) with more than $33\,Tm^3$ of known natural gas; Zagros–Mesopotamian Cretaceous–Tertiary TPS (about $14\,Tm^3$, including the North Dome/South Pars); Silurian Qusaiba (nearly $13\,Tm^3$); Arabian Sub-Basin Tuwaiq/Hanifa-Arab (nearly $8\,Tm^3$); and Amu Darya Jurassic–Cretaceous TPS with more than $6.5\,Tm^3$, including Shatlyk (Ulmishek, 2004). Subsequent discovery of Galkynysh group of fields changed that order, most likely moving up the Amu Darya system to the second place: estimates for Galkynysh have been repeatedly revised upward, the latest claimed proved and probable reserve total is as high as $26.2\,Tm^3$, the latest USGS assessment of undiscovered oil and gas resources of the Amu Darya Basin puts the total gas volume at $52\,Tm^3$ (Klett et al., 2011), and the field's total output is expected to reach $40\,Gm^3$/year in 2015.

In some instances, the tertiary migration brings some oils and gases all the way to the Earth's surface: crude oil was found oozing through rocks (in western Pennsylvania) and forming pools in many locales in the Middle East, while seepages of gas were known in antiquity by hissing sound of the escaping gas and by strange phenomena of "burning springs" or "eternal fires," most notably in what is today's northern Iraq (Kurdistan). A notable North American description of this phenomenon (a burning spring in the Kanawha River Valley of West Virginia) is worth noting as it is included in the Schedule of Property appended to George Washington's will: "The tract, of which the 125 acres is a moiety, was taken up by General Andrew Lewis and myself, for and on account of a bituminous spring which it contains, of so inflammable a nature as to burn as freely as spirits, and is nearly as difficult to extinguish" (reprinted in full in Upham, 1851, 385).

Finally, a few paragraphs on energy storage densities ($J/m^2$) of natural gas reservoirs. Methane has energy density three orders of magnitude lower than crude oil, but because of the thickness of some gas-bearing layers, storage densities of the fuel (energy content prorated per unit of surface) for the largest gas fields are similar to those of the world's largest oil fields. South Pars–North Dome field, the world's largest identified accumulation of hydrocarbons in the Persian Gulf, contains about $35\,Tm^3$ of recoverable natural gas (about 70% of its estimated total volume of $51\,Tm^3$), and the field had originally stored also about $8\,Gm^3$ of natural gas liquids (Esrafili-Dizaji et al., 2013). This adds up to about $2.1\,ZJ$ of fossil energy, and (with the area of $9,700\,km^2$) it translates to storage density of about $215\,GJ/m^2$ of sea surface.

The much smaller Urengoy (original storage of 8.25 Tm$^3$, recoverable volume of 6.3 Tm$^3$, area of 4,700 km$^2$) in Western Siberia is the world's second largest natural gas field, and its storage density is roughly 60 GJ/m$^2$ of tundra (Grace and Hart, 1991). The third largest storage, Yamburg field north of the Arctic Circle in Siberia's Tyumen region, has nearly 4 Tm$^3$ of recoverable gas and overall storage density of about 35 GJ/m$^2$ of permafrost. Analogical data for Groningen, Europe's largest onshore gas field, are 2.8 Tm$^3$ and about 110 GJ/m$^2$ (NAM (Nederlandse Aardolie Maatschappij), 2009) and 2.3 Tm$^3$ and less than 4 GJ/m$^2$ for Hugoton, America's largest natural gas field. That sprawling formation of almost 22,000 km$^2$ extends from southwestern Kansas through Oklahoma to Texas. For comparison, the world's largest reservoirs of crude oil have storage densities mostly between 200 and 500 GJ/m$^2$ (Smil, 2015).

## 2.3   RESOURCES AND THE PROGRESSION OF RESERVES

Unless we accept widespread abiogenic origin of hydrocarbons (which would allow their continuing repletion, albeit at a very slow rate), we must reckon with finiteness of natural gas stored in the uppermost layers of the Earth's crust. Extraction of all minerals, and hence of all fossil fuels, inexorably reduces the stores that were originally in place—but this process will almost never result in actual physical exhaustion of the exploited resource. In order to appreciate this fundamental reality, it is necessary to clarify the basic definitions involved in the exploitation process: different schemes have been proposed in several countries, but the one that has become the US norm is the classification used by the US Geological Survey (McKelvey, 1973; Figure 2.4).

In the United States, all publicly listed companies must file annually their natural gas reserve totals with the US Security and Exchange Commission. Those numbers are known to be rather conservative, while the figures for many OPEC nations have been exaggerated: the resources may be in place but proved reserves may be considerably smaller; this uncertainty is particularly important in Iran's case (Osgouei and Sorgun, 2012). Russia reports its reserves based on a classification adopted during the Soviet years that is not directly comparable with the Western assessments: its reporting is based solely on the analysis of geological parameters, not on an economic appraisal of reserves that can be actually recovered—but natural gas reserves in the first three

| Cumulative production | Identified resources | | | Undiscovered resources | |
|---|---|---|---|---|---|
| | Demonstrated | | Inferred | Probability range | |
| | Measured | Indicated | | Hypothetical | Speculative |
| Economic | Reserves | | Inferred reserves | | |
| Marginally economic | Marginal reserves | | Inferred marginal reserves | | |
| Sub-economic | Demonstrated subeconomic resources | | Inferred subeconomic resources | | |
| Other occurrences | Includes nonconventional and low-grade materials | | | | |

Figure 2.4  McKelvey box.

categories (A, B, and C1) are considered to be fully extractable (Novatek, 2014a).

I will explain the US classification by relying on an analogy of a large dark room whose content we proceed to examine with increasingly more powerful searchlight prior to removing its contents. The room itself is analogous to **resource base**: its ultimate size is unknown but (unreliable) estimates can be offered as to its volume or mass. The first entries into the room and the first examinations of its contents usually yield modest results: the room's volume is vast and our search light is not very powerful (sophisticate geophysical exploration was initially absent, drilling was limited to shallow depths), but we begin to get some feeling for what is in there. This knowledge is analogous to assessing a specific **resource**, that is, a concentration of materials (be they solid, liquid, or gaseous) in the Earth's crust whose economic extraction we regard as feasible, either right away or sometime in the future. In some instances, our searchlight can illuminate some large and very valuable objects (early discoveries of large hydrocarbon fields) which we promptly proceed to remove.

More specifically, resources get subdivided into economic (ready to be exploited), marginally economic, and subeconomic categories, and based on their size, we can begin to make some estimates about the further magnitude of undiscovered resources, beyond our searchlight but reasonably assumed. Further, detailed examination of viable resources moves them from subeconomic and marginal categories to the **economic reserve** category, itself subdivided into demonstrated (measured and indicated) reserves and more uncertain class of inferred reserves. Commonly used equivalents of measured and indicated reserves in other resource classifications are the categories of proved and probable reserves.

If a portion of a resource is to be classed in the category of economic reserves, it is not enough just to illuminate an object and ascertain its location and size (now done first through remote sensing and then in detail with on-site geophysical assessment). It is also necessary to probe the resource in sufficient detail (through exploratory drilling) in order to determine what techniques will be used for its recovery (e.g., horizontal vs. vertical drilling) and what cost its extraction and delivery to the market will entail. Consequently, reserves are not a given natural category but a quantity that is derived from resources through scientific, technical, and managerial means: human ingenuity keeps translating a growing share of naturally available resources into economically exploitable reserves.

This process is easily illustrated by following the progression of reserve-to-production ratios. The ratio is expressed in years, and it is simply the quotient of a specific reserve in a given year and output of a commodity during that year. Obviously, there is no economic incentive to invest into exploratory activities that would result in very large (>25 years) R/P ratios, and inversely, there can be no commitment to actual extraction unless the R/P is kept above a certain minimum (at least 10 years). In any case (and contrary to a common misperception), low or falling R/P ratios are not an inevitable sign of any imminent resource exhaustion: long-term trends of R/P ratios for the US natural gas demonstrate this reality. Between the late 1920s and the late 1940s, the ratio rose (with fluctuations) from about 20 to about 40 years, then a steep decline (until 1970, to 15 years) and a more moderate decrease brought it to just 10 years by the early 1990s, and a subsequent slow rise lifted it to 12.5 years by 2012.

This means that the ratio has been fairly steady for more than 40 years, while the US natural gas production first declined by about 20% (between 1970 and 1987) and then kept rising so that by 2011 it was nearly 10% above the 1970 mark. The ratio thus informs us primarily about the economic and technical capacities of transforming a resource to a reserve, not about any imminent peak of extraction or approaching resource exhaustion: it can be falling when low prices do not justify aggressive additions to reserves, it can be rising when new capabilities for tapping nonconventional resources expand the resource base and add rapidly to new reserves. A retrospective look shows the following natural gas R/P ratios for some of the major producing countries in the year 2000 and 2012 (with all numbers rounded to the nearest year): the United States, 9 and 13; Canada, 11 and 13; Norway, 23 and 18; Russia, 83 and 55; Algeria, 55 and 55; and Australia, 41 and 77—and the world 62 and 56 years (BP, 2014a).

In the early decades of natural gas industry, the US reserves dominated the global total: they were put at $424\,Gm^3$ in 1919, and by 1950, they were an order of magnitude higher at roughly $5.5\,Tm^3$. After reaching a peak of about $8.2\,Tm^3$ in 1970, they declined to less than $4.6\,Tm^3$ by 1994 and grew slowly for the rest of the decade, but boosted by shale gas development, they reached $5.8\,Tm^3$ by 2005 and $8.5\,Tm^3$ in 2012. In 1950, the US reserves were more than half of the global total of about $10\,Tm^3$; by 1960, North America was still the leading region with more than 40% of the world's gas reserves, but during the 1960s, it was surpassed by steadily rising Iranian total, and its reserves became a fraction of the Soviet finds.

These discoveries were made in a huge (about 2.2 million $km^2$) West Siberian Basin, mostly east of the Ob River just before it flows into the Arctic Ocean (Gazprom, 2014). They began with the Taz in 1962 and 4 years later came Urengoy field, at the time the world's largest, with the estimated ultimate recovery of $9.5\,Tm^3$ (Grace and Hart, 1991). The basin's second largest field, Yamburg (originally estimated at $4.7\,Tm^3$), was discovered in 1969, and three other extraordinary fields, Zapolyarnoye ($2.3\,Tm^3$, 1965), Bovanenkovo (almost $5\,Tm^3$, 1971), and Medvezhye ($2.4\,Tm^3$, 1967), helped to make the area the world's largest concentration of natural gas outside of the Persian Gulf (see Figure 2.2). By 1990, a year before its dissolution, the Soviet natural gas reserves reached 30% of the global total; their official total rose to about $45\,Tm^3$ in 1991. By the year 2012, the countries of the former Soviet Union held at least $55\,Tm^3$ of gas reserves or 29% of the world total, but Russia's total was, again, surpassed by Iran whose gas reserves grew by 30% during the first 12 years of the twenty-first century.

Iran's reserves are dominated by the country's share of the world's largest natural gas accumulation which contains almost 20% of all discovered gas. Qatari part (North Dome, *al-Idd al-Sharqi*) was discovered in 1971 but Iran explored the field's northern extension, South Pars in its territorial waters, only in 1990 (Figure 2.5). Although South Pars covers only about 40% of the field's area (total of 9700 $km^2$), it contains just over 70% of its proved reserves, about $25.5\,Tm^3$ of the total of $35.7\,Tm^3$ (Esrafili-Dizaji et al., 2013). With about $51\,Tm^3$ of gas reserves in place, this implies a recovery rate of 70%. The

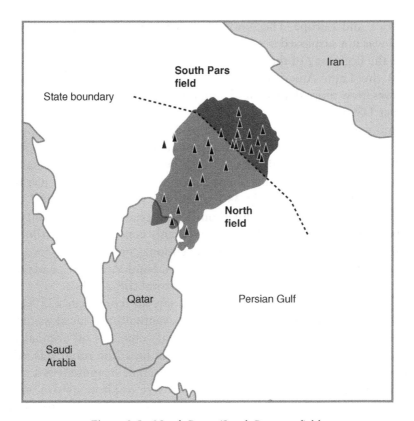

**Figure 2.5**    North Dome/South Pars gas field.

field also contains some 19 Gb (about 2.6 Gt) of recoverable condensate, and at a depth of about 1 km, there are three oil-bearing layers.

Field sizes also have a highly skewed distribution: nearly 20,000 of hydrocarbon fields have been discovered since the 1850s, but 1,087 giant fields found by 2012 hold about 72% of proved and probable oil and gas reserves (Bai and Xu, 2014). Giant fields are defined as those containing at least 500 million barrels of oil equivalent (Halbouty, 2001), be it just crude oil (that would be about 70 Mm³), mixture of oil, natural gas liquids and natural gas, or nonassociated gas (with 7.2 barrels/t and 42 GJ/t that equals roughly 3 EJ or 85 Gm³ of natural gas). Mann, Gahagan, and Gordon (2001) provide details of tectonic setting for 592 of these formations. As for the conventional reservoirs of natural gas, 10 largest fields contain more than one-third of the world's 2013 conventional gas reserves, and the share is more than 40% for the top 20 fields.

The only two regions that have not seen steadily rising natural gas reserves during half a century between 1960 and 2010 were North

America and Europe. The North American record volume reached in 1983 was not surpassed until 2011, and EU reserves have been in decline since the beginning of the twenty-first century, but in 50 years preceding 2010 the Latin American reserves had quintupled, Middle Eastern reserves grew more than 15-fold, and the reserves of the countries of the former USSR grew more than 25 times. As a result, global reserves of natural gas rose from about $19\,Tm^3$ in 1960 to $72\,Tm^3$ in 1980, to at least $140\,Tm^3$ in the year 2000, and to more than $177\,Tm^3$ in 2010, more than a ninefold rise in 50 years. As with most energy resources, natural gas endowment has a highly skewed spatial distribution, with just four countries (Iran with roughly $34\,Tm^3$, Russia with $33\,Tm^3$, Qatar with $25\,Tm^3$, and Turkmenistan with about $18\,Tm^3$) harboring nearly 60% of global reserves in 2012, a slightly more highly concentrated distribution than that for crude oil: when nonconventional proved oil reserves are included, the top four countries, Venezuela, Saudi Arabia, Canada, and Iran, account for about 56% of the world total (BP, 2014a).

To know (within an acceptable error, say, no more than ±25%) the total volume of natural gas that could be eventually produced would be much more valuable than to know the total volume of oil originally in place. Of course, we know that the resource base (the room, gas originally in place) is finite, but to ascertain its ultimate magnitude is a notoriously difficult challenge, and even if we were to know its size with a high degree of accuracy (after our searchlights will have penetrated the furthermost recesses of the room), we still could not immediately translate this knowledge into clear assessment of ultimate production, the sum of already produced gas, remaining reserves, conventional reserve growth (the overall volume for a given location tends to increase with time as more exploration is done), and undiscovered conventional gas.

Because reserves are determined by our technical abilities and economic possibilities, what appears unattainable today can become a matter of routine recovery in a few years or decades: recent rapid expansion of natural gas production from American shales is a perfect example of this reality as a combination of newly affordable horizontal drilling and improved hydraulic fracturing transformed a significant share of a previously untapped resource into highly economical reserve of gaseous fuel. But as these nonconventional resources have not been a part of the past global assessments, I will leave them in the following comparisons and do their appraisal in Chapter 6.

Having a good estimate of ultimately recoverable natural gas would allow major producers to do better long-range planning on local, regional, or national basis, and it would allow us to make better assessments of long-term global energy supply prospects of modern civilization. Here is what we know. The first global assessment of ultimately producible natural gas was done by using a rough multiplier relating gas to ultimately recoverable crude oil. Half a century ago, such crude estimates yielded at least $216\,Tm^3$, and in the mid-1970s, the most likely volume was put at $280\,Tm^3$ (Hubbert, 1978). A decade later, the USGS estimate was nearly $260\,Tm^3$, and then regular global USGS assessments, based on evaluating 171 geologic provinces, raised the volumes to about $295\,Tm^3$ for 1990 and $435\,Tm^3$ for the year 2000 (USGS (United States Geological Survey), 2000).

The year 2000 total was composed of roughly $50\,Tm^3$ of cumulative production (less than 12% of ultimately recoverable reserves, with just below half of that total extracted in the United States), $135\,Tm^3$ of remaining reserves, $104\,Tm^3$ of reserve growth, and $146\,Tm^3$ of undiscovered conventional gas. In 2012, the USGS put its latest estimate of undiscovered conventional natural gas outside the United States about 9% higher, at $159\,Tm^3$ (Schenk, 2012). And the USGS had also estimated that in 2012, reserve growth, that is, potential additions to conventional gas resources of the world (again outside the United States) in discovered giant gas fields (with reported gas in place of at least $85\,Gm^3$), was at least $40\,Tm^3$ of gas and $16\,Gb$ (about $2.2\,Gt$) of gas liquids (Klett et al., 2012).

But this ascending progression of reserves does not mean that eventually all undiscovered but potentially producible resources will shift into reserve category—and it would be quite misleading to think that the story of exploitation will end with physical exhaustion. Significant shares of such resources will always remain uneconomical (in too small concentrations, too scattered, of exceedingly poor quality), and in the case of hydrocarbons embedded in reservoir rocks, only the actual mining of those (more or less) porous formations (as is done with Alberta with oil shales) could recover virtually all fuels originally present in place. This means that in virtually all cases, conventional extraction of underground reservoirs leaves behind significant shares (as much as 70–75%) of liquids and gases inside the porous rocks.

A more practical indicator of extraction prospects is the cost of marginal production, a changing sum that is affected both by technical advances and by the ability of economies to pay the price. Technical advances may lower the cost of existing extraction or they may increase it but make available previously unexploitable resources and hence improve the security of supply that may be worth of higher costs. Affordability varies among economies, but there is no simple positive correlation between the level of per capita GDP and the ability to pay more (still relatively poor China has huge foreign reserves and the country has been a rising importer of all fossil fuels), and some countries may have little choice paying exorbitant prices (the case of Japan's LNG imports after Fukushima: for more, see Chapter 6).

Once the marginal costs become unsupportable, a particular extraction locality (and eventually all, or virtually all, production of a specific resource in a country) is gradually abandoned (British underground mining is an excellent example of this reality)—and economies respond by a mixture of reduced uses and resource substitutes. Hence, there are no output peaks followed by precipitous collapses, rather (slower or faster) shifts toward new use and supply patterns. The cited assessments indicate that global natural gas extraction can keep rising to new record levels for decades to come. Assuming that the recovery of additional $400\,Tm^3$ of conventional gas will follow the normal (Gaussian) function roughly fitted to the past progression, the peak of global extraction would come around 2050, and after gradually declining, the annual output at the beginning of the twenty-second century would be still about as large as it was in the year 2000.

But production curves of natural resources are perfectly symmetrical only in theoretical studies; in the real world, they show all kinds of deviations, and their progression is dependent on inherently inaccurate estimates of ultimately recoverable fuel totals: hence, a less steep progression of global gas extraction would bring a production peak perhaps only around the year 2070—and various assumptions regarding the recovery rates of nonconventional natural gas resources (see Chapter 6) could raise the output peak as well as lengthen the duration of natural gas era. Here is perhaps the best illustration why such long-range forecasts are little more than amusing exercises with little relation to reality.

In 1956, M. King Hubbert, American geophysicist and advocate of symmetrical production curves for mineral resources that appear to be well suited to forecast output peaks, put America's ultimately recoverable natural gas resources at $24\,Tm^3$ (Hubbert, 1956), and in 1978, he raised that total to $31.2\,Tm^3$ (Hubbert, 1978). These totals led him to forecast the US peak gas production in 1973, either at about $400\,Gm^3$ (with the 1956 total) or $620\,Gm^3$ (with the 1978 total). But by the end of 2013, the cumulative US gas extraction reached $34\,Tm^3$, nearly 10% above Hubbert's grand total of all recoverable gas, and the 2013 annual production was about 12% above the previous (1973) record.

And while the biennial assessments of technically recoverable US natural gas endowment were reporting a fairly steady volume between 1990 and

2004 (rising from 28.3 to 31.7 Tm³), their totals rose to 53.7 Tm³ in 2010, and the latest assessments by the Potential Gas Committee (2013) put the remaining potential gas resources (conventional, tight, shale, and coal bed) at 67.5 Tm³ (with shale gas accounting for 45% of that total or more than the conventional resources in the year 2000), which is more than twice as high as Hubbert's higher (1978) value: so much for orderly cumulative production curves and pinpointed peaks.

Similar uncertainties regarding ultimate recoveries and the most likely production peaks apply to crude oil, and hence, it is only as an interesting aside that I introduce some published comparisons of the two resources. Laherrère (2000) put the ultimately recoverable gas at 1.68 Tb of oil equivalent, the total nearly identical to his estimate of ultimately recoverable crude oil. In the same year, World Petroleum Assessment put the total volume of recoverable conventional oil nearly 60% higher at 2.659 Tb compared to 2.249 Tb of oil equivalent in natural gas. That is only about 18% difference, most likely within the estimate's error range, and hence it is a fair statement, as the first order of approximation, that our best understanding of recoverable hydrocarbon resources puts crude oil and natural gas at nearly the same total energy level.

But the combination of more focused appraisals of unconventional resources and technical advances in their recovery may eventually make the marketable reserves of natural gas significantly higher than those of liquid fuels. A recent Canadian gas resource assessment illustrates this continuing process. In November 2013, the National Energy Board released the first study of marketable unconventional petroleum resources in the Montney Formation in west-central Alberta and northeastern British Columbia (NEB, 2013). Montney siltstones and sandstones are estimated to hold 12.7 Tm³ of marketable natural gas, 2.3 Tm³ of marketable natural gas liquids, and 1.1 Gb of marketable crude oil. These reserves are equivalent to nearly half of the reserves in the world's largest gas field in the Persian Gulf, and the Board concluded that the overall volume of natural gas resources in Western Canada is likely to increase as other unconventional formations receive closer attention. I will take another look at nonconventional resources of crude oil and natural gas when I will compare their global endowment in Chapter 6.

# 3

# Extraction, Processing, Transportation, and Sales

As long as most of natural gas production was associated with crude oil extraction (i.e., for most of the twentieth century), the term oil and gas industry was a much more apt description of activities involved in finding, producing, and transporting gaseous fuel, and of course, in many hydrocarbon basins, it is still the case. But numerous post-WWII discoveries of nonassociated (dry or wet) gas; development of new giant and supergiant gas fields in North America, Europe, Asia, Africa, and Australia (including, in all regions, important offshore deposits); construction of transcontinental pipelines; and emergence of large-scale liquefied natural gas (LNG) trade have created a distinct, and now truly world-spanning, natural gas industry that now expects further vigorous growth in decades ahead.

This chapter presents a basic how-we-do-it sequence of activities that make up this quintessential (yet largely hidden) modern industry: how we look for new, commercially viable resources; how we extract the gas from reservoirs, coal beds, and shales; how we process it before we send it into pipelines; how we transport it across long distances and then distribute it to individual consumers; how we store it for often prolonged periods of high demand; and how, and how much, we have been selling to consumers. The only segment of modern natural gas industry that is deliberately left out is the intercontinental transportation of LNG, an increasingly important activity that will get a closer look in the fifth chapter dealing with the emergence of the global gas trade. But before I

*Natural Gas: Fuel for the 21st Century*, First Edition. Vaclav Smil.
© 2015 John Wiley & Sons, Ltd. Published 2015 by John Wiley & Sons, Ltd.

begin a brief systematic review of exploration, extraction, and transportation activities, here is perhaps the most apposite place for a few paragraphs about the early history of natural gas production and the reasons for its relatively slow pre-WWII development.

Some histories of natural gas cite spurious claims of very early use in China, but properly documented exploitation in Sichuan, the country's most populous landlocked province, goes back only to the Han dynasty at about 200 BCE (Needham, 1964). Percussion drills (heavy iron bits at the end of bamboo cables raised by men rhythmically jumping on levers) were used to dig relatively deep wells, and the gas, transported by bamboo pipes, was burned to evaporate brines in large metal pans, an ingenious way to produce salt in the landlocked province hundreds of km from the closest sea coast (Figure 3.1). Eventually, a small share of the gas was also used for lighting and cooking. This industry continued for more than two millennia, with the bores reaching a depth of 150 m by the tenth century, and the record Xinhai well was 1 km deep in 1835 (Vogel, 1993).

There are no other documented instances of such gas production, and even after substantial volumes of associated gas became available with

Figure 3.1   Chinese percussion rig. Reproduced from Song 1673.

the development of new oil fields during the latter half of the nineteenth century, most of the gas was simply flared, a wasteful (and environmentally undesirable) practice that (as I will explain in some detail in Chapter 7) has continued for too long in too many places in Asia and Africa and that is still taking place today even in North America. The first use of natural gas in the United States actually predates the first commercial exploitation of crude oil (in 1859 in Pennsylvania): the first shallow (only about 8 m) gas well was dug in order to expand the flow from a natural seepage in 1821 in New York (Fredonia near the Lake Erie) by William A. Hart (Castaneda, 2004). The town, and later a nearby lighthouse on Lake Erie, received gas through hollowed-out pine-log pipes.

Percussion drilling was speeded up in the age of steam, and Edwin L. Drake used a small engine to complete America's first commercial oil well (just 21 m deep) at Oil Creek in Pennsylvania on August 27, 1859 (Brantly, 1971). But the great pioneering era of oil extraction during the late nineteenth century had no counterpart in large-scale development of gas industry. Three main factors explain this absence: ready supply of cheap coal and newly abundant refined liquid fuels; technical limits, above all the absence of inexpensive seamless pipes able to withstand higher pressure and reliable compressors to propel the gas over long distances (and hence decades of comments about large volumes of stranded gas and discovered but undeveloped resource because of no access to markets); and ubiquitous availability of coal (town) gas in all major and medium-sized cities.

The United States was the only country where a sizable natural gas industry began to be developed during the 1920s, but as with most other segments of the economy, the Great Depression slowed down (but did not reverse) the process. Demand for industrial production during WWII brought a nearly 50% increase in the American natural gas extraction between 1940 and 1945, but the greatest period of expansion came only after WWII. The industry was driven by the need to supply natural gas for expanding cities (and suburbia) and industries as natural gas became an essential energizer of new economic prosperity. The US natural gas production grew 2.3 times during the first postwar decade, and then it doubled between 1955 and 1970. Another sign of its rising importance was the fact that in the late 1950s its value (including natural gas liquids (NGLs)) surpassed the value of coal production (USCB [US Census Bureau], 1975).

Europe began to shift to natural gas slowly in the 1960s (following gas discoveries in the Netherlands and in the North Sea), and the consumption accelerated during the 1970s as the North Sea gas reached the continent's markets in rising volumes and during the 1980s when the Soviet lines of unprecedented lengths and capacity brought the Western Siberian gas all the way to Venice, Vienna, and Berlin (for details see Chapter 5). Japan has to import all of its gas, and it began to do so in the late 1960s and has been increasing its dependence on imported LNG ever since (and at an even faster rate after the post-Fukushima closure of its nuclear reactors). China has seen the most recent period of substantial production increases and extensive construction of long-distance pipelines: since the year 2000, it has brought its own gas from its westernmost province (Xinjiang), and the imported gas from the former Soviet Central Asia, all the way to its large eastern coastal cities.

## 3.1    EXPLORATION, EXTRACTION, AND PROCESSING

Combination of a large amount of practical field experience (accumulated during more than 150 years of hydrocarbon industries) and considerable scientific and engineering research (in areas ranging from fundamental tectonic geology and 3D, even 4D, simulations of oil and gas reservoirs to tools and procedures for remote sensing, horizontal drilling, and gas processing) has transformed every step of modern natural gas development into a highly complex and invariably computerized endeavor aimed at minimizing risks and maximizing financial returns. This has created a massive supply of clean and flexible fuel at affordable prices, but the almost universal presence of computer controls and high levels of mechanization and automation have resulted in only limited labor opportunities.

Detailed US labor statistics divide the total employment in oil and gas industry (there is no separate accounting just for natural gas extraction) into three categories—drilling, extraction, and support, and at the end of 2012, drilling employed more than 90,000 people, and extraction 193,000, and there were 286,000 support jobs, altogether about 570,000 people or an equivalent of 0.5% of total US private sector employment (USEIA [US Energy Information Administration], 2013a). And even the most liberally calculated total of new direct and indirect jobs created by the recent rapid expansion of oil and gas production from shales is, unfortunately, just a small fraction of jobs lost in the US manufacturing since the year 2000 (Smil, 2013a).

## 3.1.1   Exploration and Drilling

Seismic surveys rely on sound waves (generated at the surface by truck-mounted vibration pads or dynamite charges, in the ocean by firing compressed air from air guns towed behind a vessel) reflected from rock formations and intercepted by sensitive receivers (geophones or hydrophones) spaced along receiver lines that are usually 300–600 m apart (Li, 2014). In 1928, John C. Karcher's Geophysical Research Corporation drilled the first oil-producing well pinpointed by reflection seismography (in 1940, Karcher's company was renamed Texas Instruments, and after WWII, it emerged as one of the leaders in microelectronic revolution).

3D seismic imaging was invented by Humble Oil in 1963, developed by Exxon in the Friendswood oil field near Houston during the late 1960s, and it had rapidly replaced 2D seismic reflection as a key tool for decision-making in hydrocarbon drilling (Ortwein, 2013). This advance, requiring large amount of collected data, would have been impossible without processing capacities made available by the development of semiconductors and, starting in the early 1970s, of microprocessors. Acquired data are processed by computers to generate 3D visualizations (including large-scale immersive displays where people can walk inside imagery) of subsurface formations that help to envisage the best way of drilling and reservoir exploitation.

The need for ever better exploratory techniques is made clear by the cost of drilling. Large rigs are commonly leased at daily rates of $200,000–$700,000, and the average onshore exploratory well in North America costs $4–5 million. US historical data show the average cost of natural gas wells increasing from about $500,000 in the mid-1980s to $1 million by 2002 and then rising rapidly to $1.5 million by 2005 and $3.9 million by 2007 when the series ends (USEIA, 2014a). Drilling, fracking, and completing a Marcellus shale gas well averaged $7.65 million in 2011, with high-volume hydraulic fracturing ($2.5 million) and land acquisition ($2.2 million) being the two highest expenses, followed by horizontal and vertical drilling at, respectively, $1.21 million and $663,000 (Hefley et al., 2011). Wells in the interior of South American, African, and Asian countries (in some cases without road access, necessitating transport by helicopters) could be easily two or three times as expensive, offshore wells usually require on the order of $20–40 million, and those in deepwaters need up to $100 million.

All early drilling was done by the ancient percussion method widely used in premodern China. The method appeared primitive: as already noted, heavy iron bits, fastened to bamboo cables attached to derricks and levers (also made of bamboo), were repeatedly raised and dropped (originally powered by men jumping on the levers) to the bottom of a hole, but the process could eventually produce surprisingly deep wells. Early development of US and Russian hydrocarbon industries used sturdy wooden derricks and manila ropes and it was powered by steam engines; steel cables and diesel engines came later, and cable tool rigs were still in common use by the middle of the twentieth century, 50 years after the introduction of rotary drills (Smil, 2006).

These superior tools, patented in 1909 by Howard Robard Hughes, consisted of two rollers arranged at an angle to each other as they rotated on stationary spindles while the entire bit rotated at the end of tubular drilling string and advance through rocks 10 times as fast as the percussion tools. Hughes Tool Company (now operating as Hughes Christensen, belonging to Baker Hughes) further strengthened its leading position by patenting its tricone bit in 1933, still the dominant tool in modern drilling. Rotary drills have either embedded industrial diamonds or synthetic diamonds bonded to tungsten carbide. Above the drill bits advancing into rock formation is a massive, complex structure of supports and prime movers needed to bear the weight of a drill string and to impart the rotary motion.

First wells in previously unexplored areas are called wildcats, a reference to possible mishaps while encountering unexpected high reservoir pressures that can result in blowouts. Once the presence of hydrocarbons is confirmed by exploratory drilling, the same type of equipment and the same procedures are used to drill production wells that are appropriately spaced in order to optimize field productivity, and in the later stage of exploitation, also injection wells are used to introduce water or gases into a reservoir: these secondary recovery methods can significantly boost the shares of oil and gas eventually recovered from a reservoir. I will note only the essentials of the drilling process that combines power and control but remains challenging and risky: detailed descriptions of evolving techniques and operating modes are available in Lewis (1961), Brantly (1971), Davis (1995), Horton (1995), Selley (1997), Smil (2006, 2008), and Devold (2013).

Once a site is pinpointed by geophysical exploration, it is prepared for drilling: if need be, vegetation is cleared and topsoil is removed, surface hole is drilled and surface casing is inserted and cemented into position (it isolates the bore from its surroundings and prevents groundwater contamination), and a blowout preventer (safety valves able to withstand high pressure and prevent any sudden eruption of hydrocarbons from the bore) is installed. Drilling rigs are visible from afar as steep, narrow conical steel structures. These derricks are topped with the crown block consisting of pulleys and traveling blocks able to add and remove drill pipe sections (about 10 m long) and support mass of the suspended drill string, connected drill pipes, and bottom-hole assembly with conical drill bit (Figure 3.2).

Diesel engines are used to power a rotary table that turns the drill string clockwise. As a well goes deeper, the weight of drill string would quickly surpass the total of 6.5–9 t, typically the mass that produces the fastest rate of rock penetration: fewer than 50 sections (reaching to 300 m below the surface) would weigh that much, and that is why the crown block assembly at the top of the derrick (whose mechanical advantage enables it to support several hundreds of tons) is needed to keep only the optimum mass pressing down on a bit. Powerful pumps are also needed to force down the drilling fluid into the well and then to remove it.

Pressurized drilling fluids are commonly called drilling muds, a misnomer for a complex mixture of liquids, solids, and gases designed to remove suspended cuttings as they are drawn upward through the space between drill string and wellbore walls; muds also clean and cool the rotating bit, inhibit its corrosion, and help to prevent cave-ins. Blowout

Figure 3.2   Modern drilling rig. © Corbis.

prevention (subsurface safety valve) and an arrangement for monitoring the progress of drilling are also essential. Drilling progress depends on the hardness of the rock, ranging from no more than 1 m/h in granite to about 20 m/h in chalks and soft sandstones; daily advance on the order of 100 m is common. Deepest hydrocarbon wells were about 1.5 km during the 1920s, more than 4 km by the late 1930s, and 6 km by 1950, and 9 km wells were drilled in Oklahoma's Anadarko Basin during the 1970s. But average depths of US natural gas wells rose from just over 1 km in the early 1950s to about 2 km by 2010, although in some basins depths of 3–4 km are common.

As the drilling proceeds, new casing sections (threaded steel pipes) are inserted and cemented into position, and the progress of a wellbore is monitored (logged) to ensure that it proceeds in the planned direction and that its integrity is safe. Both cementing and logging of wells became common after WWI, and the two innovating companies still dominate their respective fields nearly a century later. In 1922, Erle P. Halliburton patented a cement jet mixer, and his Oklahoma company grew into a leading worldwide provider of oil field services including not just cementing but also drill bits, well completion, pipe perforation, well testing, and control of accidental blowouts.

Electric well logging, pioneered by Conrad Schlumberger (at the French École des Mines) in 1911, became an essential part of exploratory drilling. In 1927, Schlumberger's company introduced the first electrical resistivity log to record successive resistivity readings; in 1931, it began to exploit the spontaneous potential that arises between an electrode in the borehole fluid and water in permeable beds through which the bore advances; and in 1949, it introduced induction logging to measure resistivity after sending alternating current through rock formations. Blowout preventer is yet another key innovation going back to the 1920s: James S. Abercrombie and Harry S. Cameron introduce it in 1922, and their first designs could control pressures of up to 20 MPa, while today's devices can contain up to 100 MPa in wellbores tapping deep formations.

Combination of these techniques could distinguish between permeable and impermeable layers and between hydrocarbon-bearing and water-bearing substrates. Schlumberger has remained a well-logging leader even as new techniques and microelectronics transformed the well monitoring process. Small, slim tools at the end of flexible wires that are lowered into boreholes now log different gamma-ray signatures of sandstones and shales, while neutron logs help to assess rock porosity. Logging used to be done after a well was completed, but new sensors embedded in bottom-hole assembly allow for logging while drilling. Horizontal drilling—whose obvious advantage is the ability to penetrate a much large volume of reservoir rock—used to be highly demanding and expensive, but technical advances have turned it into widely affordable and routinely accomplished operation: I will describe its progress in Chapter 6 when dealing with the expansion of the US shale gas extraction.

A new beginning in the history of oil and gas exploration came in November 1947 when the first drilling rig operating out of sight of land discovered oil 16 km off the Louisiana coast. Onshore production can come from wells with low or marginal productivity whose operation would not be justified in offshore wells whose cost is considerably higher. Production in shallow waters (up to about 100 m) can be done from fixed platform with seabed foundations; in deeper waters, options include gravity bases (giant concrete structure resting on sea bottom and supporting operating deck), tension-leg platforms (held in place by tensioned cables), semisubmersible platforms or floating production, storage, and off-loading structures. Advances in deep-sea construction made it possible to build complex subsea production systems on the seabed with multiple wellheads tied to a pipeline to a shore.

## 3.1.2 Well Completion and Production

When the drilling is completed, different types of steel casing are installed and cemented in place: they are designed to support years of accident-free production by insuring free flow of gas and by preventing ingress of other gases or liquids as well as seepage of gas and condensates into overburden rocks and, a most unwelcome occurrence, into aquifers used to supply drinking water. The uppermost segment is the widest but relatively short (10–15 m) conductor casing followed by narrower surface casing that usually goes down a few hundred meters through the aquifer(s) zone. Intermediate casing fills most of the wellbore penetrating through the overburden rocks, and the production casing running through the reservoir rock completes the installation.

When the casing is in place, a smaller pipe, production tubing with diameters of mostly 5–25 cm, is inserted into it and packed in position. At that point, no gas or liquid could flow through the production tubing as it is separated from the reservoir rock by its wall, casing wall, and cement. Bullet perforators were used to make holes in casing and in cement, but now the entry is achieved by jet perforation with electrically ignited charges making small holes through the casing and cement and into the surrounding rock to allow the entry of natural gas. In formations containing loose sand, these perforations must be protected by screen, or else the tubing would fill with sand.

Wells are completed by installing permanent wellheads designed to control and monitor the gas flows, and in order to prevent leakage and blowouts, they are designed to withstand pressures of up to 140 MPa, or about 1,400 times higher than the normal atmospheric pressure of 101.3 kPa. This is done by a complex assembly of casing and tubing heads and a series of valves (commonly called Christmas tree). This structure is usually about 1.8 m tall, and it houses tubes and valves used to monitor the product flow; master gate valve matches the diameter of production tubing and is normally left open; pressure gauge is placed right above it, followed by wing valve, swab valve, and variable choke valve.

Gas flow depends mainly on reservoir characteristics (above all on the pressure), bore diameter, and age of the well. Daily productivities per well range over four orders of magnitude: they may be as high as $10^6 m^3$ and as low as $10^2 m^3$. A low-productivity well in Saskatchewan now yields less than $500 m^3$/day, and the Canadian average in 2012, with about 145,000 operating wells producing $167 Gm^3$ of gas (CAPP [Canadian Association of Petroleum Producers], 2014), was about $3100 m^3$/day ($1.1 Mm^3$/year). The US mean for the same year, with about 480,000 operating wells producing roughly $681 Gm^3$ of dry gas (USEIA, 2014b), was $3900 m^3$/day. In contrast, new wells in West Siberia's Urengoy giant field average 500,000–$580,000 m^3$/well a day (Seele, 2012), and 30 wells drilled in the North Dome to supply Qatargas 2 LNG project average $2.75 Mm^3$ of wet gas a day, while the daily average for 22 production wells supplying Qatargas 1 is $2 Mm^3$/well, and for 33 wells of Qatargas 3, it is $1.2 Mm^3$/well (Qatargas, 2014a).

Secular decline of average well productivity is well illustrated by reliable US and Canadian data. In the United States, average well productivity rose from less than $5,000 m^3$/day during the 1950s to the peak of $12,300 m^3$/day in 1971, and by 1985, it declined to $4,500 m^3$/day (Schenk and Pollastro, 2002). Afterward, it kept on declining at a much slower rate to less than $3,300 m^3$/day by 2005, and it rose slightly to almost $3,900 m^3$/well by 2012 (Figure 3.3). Both the absolute rates and the pace of decline have been similar for Canadian natural gas extraction: in 1980, the average was $13,500 m^3$/well, and now, it is just around $3,000 m^3$/well (CAPP, 2014). Extraction can be also deliberately restricted to prolong a field's role as a major production: Groningen's output per well

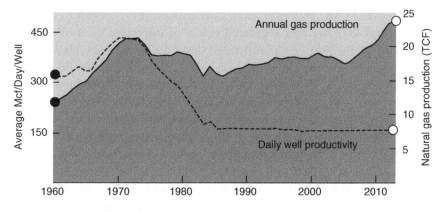

**Figure 3.3**   Productivity of US gas wells.

peaked in 1976 at 814,000 m³/day and declined afterward, and between 2010 and 2020, it is capped (with 296 wells and 43.6 Gm³/year) at 400,000 m³/well a day (NL Oil and Gas Portal, 2014). Declining production translates into falling revenues over time unless, of course, there is an intervening substantial price rise.

Detailed American data also show that slightly more than 10% of produced gas does not leave hydrocarbon fields (USEIA, 2014a). In 2012, gross withdrawals amounted to 836 Gm³ (43% from gas wells, 17% from oil wells, and the rest from shale and coal bed wells), and while only about 0.7% of all withdrawn gas was vented or flared, 11% were used to repressurize hydrocarbon fields. Once reservoirs begin losing their natural pressure, two major techniques are used to displace remaining liquids toward wellbores and extend the recovery span: flooding with water and injection of gases. Water (fresh, brackish, or ocean) is usually injected into production zones while gas is introduced into reservoir caps to maintain pressure on the underlying liquid. Any readily available gas would do, and some fields use $CO_2$ (an option that has been also promoted as a minor form of future carbon sequestration), but the use of coproduced gas is a convenient, and rewarding, choice because it may raise the overall recovery by 20–40%.

American natural gas wellhead prices (charged by producers selling natural gas into the interstate market) have been always expressed in dollars per thousand cubic feet, abbreviated as $/Mcf (because in the US usage M stands for 1,000, not for 1,000,000 as in the International System of Units). Historical record of average wellhead price goes to 1922 when it was (all prices in current dollars) $0.11/Mcf (USEIA, 2014c). During the crisis years of the 1930s, the price fell to $0.05/Mcf, and since 1938, the Federal Power Commission (FPC) began to enforce the Natural Gas Act that regulated prices of interstate sales of natural gas. In 1954, the Supreme Court ruled that all natural gas sales are subject to the FPC regulation, and hence, all wellhead prices would be regulated, with prices set high enough to cover the cost of production and a fair profit (Foss, 2004; NaturalGas.org, 2014). This system held prices too low, and it created an enormous bureaucratic burden that was not resolved by several attempts at reform. Low prices encouraged higher consumption but discouraged exploration and created supply shortage.

This was finally addressed by the Natural Gas Policy Act of 1978 that created a single national natural gas market and allowed market to set the wellhead price of natural gas. Wellhead price, whose average was $0.16/Mcf during the 1960s and close to $0.20 during the early 1970s, rose to $1.18/Mcf by 1979. Deregulation of US natural gas prices began in 1985, and it was formally accomplished with the Natural Gas Wellhead Decontrol Act (NGWDA) in 1989, with all the remaining regulations on wellhead sales to be removed as of January 1, 1993. Finally, in 1992, a federal order completed the unbundling transportation and sale services: pipelines could no longer sell gas sales, leaving customers a choice of selecting sales, transportation, and storage from any provider. During the 1980s and 1990s, wellhead prices fluctuated mostly between $1.5 and 2.5/Mcf and then climbed steeply to the annual record of $7.97/Mcf in 2008 before falling to $2.66/Mcf by 2012 (Figure 3.4).

Finally, a few paragraphs are presented on the power density of natural gas extraction. Both the areas of individual well sites and their density in major gas fields vary greatly. For example, Jordaan, Keith, and Stelfox (2009) found that gas wells in Alberta claim as little as 1.5 and as much as 15 ha (average of 3 ha/well), and while typical well density in conventional US gas fields is just 0.4 wells/km², the rate is 1.1–1.5 wells/km² in Barnett shale, and in some shale gas developments, the density will be up

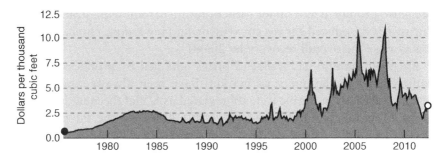

Figure 3.4   US natural gas wellhead prices.

to 6 wells/km$^2$ (NYSDEC [New York State Department of Environmental Conservation], 2009). But in general, future well densities should decline because multiple horizontal wells can be drilled more economically from a single well pad: in Marcellus shale, a vertical well may be exposed to no more than 15 m of the reservoir layer, while a horizontal well can reach 600–2,000 m within the targeted formation (Arthur and Cornue, 2010).

Power densities of natural gas production (gross energy flux per unit of surface claimed by wells and associated infrastructure for gathering and field processing of gases) are commonly around 2,000 W/m$^2$ and can be an order of magnitude higher in supergiant fields. Perhaps, no other giant natural gas field is as inconspicuous as Groningen: its extraction proceeds at 29 remotely controlled production clusters, each one with 8–12 wells that are arranged in a strip and connected to associated gas treatment plants (NAM [Nederlandse Aardolie Maatschappij], 2009). Each cluster occupies just 11 ha for the total of about 300 ha, and with annual 2010–2020 production capped at 43.6 Gm$^3$ of annual output, this puts the field's power density of extraction at roughly 16,000 W/m$^2$ (Figure 3.5)

**Figure 3.5** A well cluster (one of 29) in Groningen gas field. Imagery © 2014, Aerodata International Surveys, Map Data © 2014.

### 3.1.3   Natural Gas Processing

No processing would be required if raw natural gas were a pure, or virtually pure, methane with a negligible admixture of $N_2$ and $CO_2$. But such natural gases are exceedingly rare, and as already explained, mixtures of gases and liquids that come out of the ground are far too heterogeneous and contain at least two and often half a dozen compounds that should not be present in gas that is transported by pipelines and burned for a variety of industrial, commercial, and household uses. Processing of raw natural gas ensures the delivery of product that is not either pure $CH_4$ or a perfectly standardized homogeneous mixture of allowed components but a fuel (or feedstock) whose composition meets a variety of prescribed conditions and limits.

Processing of natural gas is far less challenging than the processing of crude oil: refining and production of liquid fuels involve high temperatures, high pressures, and catalytic reactions. In contrast, gas processing is largely a matter of separation, of removing liquids and undesirable (corrosive or incombustible) gases in order to meet quality standards for pipeline transportation and commercial use. These standards prescribe maximum concentrations of constituents other than methane: limits are set for $CO_2$, $N_2$, $O_2$, and water vapor present in the gas, and restrictions on the energy density of transported gas also require the removal of NGLs (condensates), mostly ethane and propane.

Some simple field processing is done right at or near producing wellheads: gravity separation in horizontal vessels (with water settling at the bottom, liquid hydrocarbons above it, and gases on the top) is the simplest and most common field option, and the first oil/gas separator was in operation already in 1863. But in large gas fields, processing has become a major industry, and raw gas from many wells is led through (small-diameter, low-pressure) gathering pipelines to processing plants. Depending on its composition, the raw gas will be subject to at least two and often all four of these purification processes: removal of oil and condensate, separation of valuable NGL, removal of water, and removal of $CO_2$, any sulfur compounds, and, sometimes, $N_2$ (Mallinson, 2004; IHRDC [International Human Resources Development Corporation], 2014; NaturalGas.org, 2014).

Gas must be dehydrated in order to prevent water condensation and potential corrosion of pipes. This is done mostly by absorption by liquids but adsorption by solids is also used. Liquid desiccants used for absorbing water are either diethylene or triethylene glycol. When in contact with raw gas stream, these hydrophilic compounds will absorb water, and as they get heavier, they sink to the bottom of a contact chamber, and water is then removed by evaporation (glycols have a much higher boiling point, 288°C, than water) and the desiccants are reused. Wet gas is introduced at the bottom of vertical cylindrical (5–8 m tall) columns (contactors), it ascends through glycol solution flowing from the contactor's top (trays or packing materials are used in order to maximize the contact surface), and dry gas leaves at the top of a contactor.

Water can be also removed by adsorption onto the surfaces of solid desiccants, usually alumina or silica gels placed in adsorption towers. This process is both highly efficient and suited for large volumes of highly pressurized gases and that is why it is used in so-called straddle systems, dehydrators placed on gas pipelines downstream of compressor stations. Dehydration uses two parallel columns in order to allow for the regeneration of the adsorption medium in the column saturated with water. Adsorption is more expensive than glycol absorption, but it is much more effective (it can reach dew points as low as −100°C).

Pipeline gas should be also devoid of any gases whose presence could lead to formation of acids: this requires removal of $CO_2$ and $H_2S$ (notable for its "sour" rotten smell) as well as of other sulfur compounds ($CS_2$, COS, mercaptans). The principal process used to remove $H_2S$ ("sweetening" the gas) is absorption by amine solutions in a manner similar to glycol dehydration (gas ascending through columns containing either mono- or diethanolamine). Removal by solid adsorbents is also possible, and sulfur is usually recovered by the Claus process (a sequence of oxidation, cooling, reheating, catalytic conversion, and cooling) that produces elemental S to be used in the synthesis of sulfuric acid. Large yellow heaps of the element seen in the Port of Vancouver, ready to be exported to Asia, come from the desulfurization of Western Canada's sour gases high in $H_2S$ (Figure 3.6).

Separation of liquid hydrocarbons is an obvious necessity for gases associated with crude oil: some of them have even higher shares of heavier alkanes than nonassociated wet gas rich in higher alkanes. Decreased pressure may accomplish this separation as soon as the mixture of gases and liquids reaches the surface, or the process may be helped simply by gravity as gases and liquids stratify over time inside enclosed tanks. But hydrocarbon recovery and fractionation are needed even for dry gas dominated by methane. There are three main reasons for this: to reduce hydrocarbon dew point and hence to eliminate condensation during the transportation of the gas, to meet the requirements for the prescribed heating value of transported gas, and to separate

**Figure 3.6**   Sulfur in the Port of Vancouver. © Corbis.

NGLs and market them as valuable feedstocks (ethane, propane, butane) and fuels (propane, butane).

Low-temperature separators are used to remove light oils and NGLs by taking advantage of different boiling points of alkanes: at atmospheric pressure, methane boils at –161.5°C, but ethane can be separated at –88.6°C, propane at –42°C, and isobutane at just –11.7°C. Removal of light alkanes is done by cryogenic expander process. Pressurized rich (or wet) gas is first precooled by a counterflow of cold methane to about –34°C, and this liquefies all $C_{3+}$ alkanes; propane ($C_3H_8$) and butane ($C_4H_{10}$) are usually marketed as liquefied petroleum gases (LPG) and distributed in refrigerated tank trucks and ships and stored in recognizable large industrial bullet-like tanks and small portable containers to be used as heating and cooking fuels. Both alkanes can be also sold separately as valuable feedstocks (see the next chapter for details). Pentane ($C_5H_{12}$) and any traces of heavier molecules are liquids that are mostly blended into gasolines.

This leaves the mixture of methane $CH_4$ and $C_2H_6$ that enters demethanizer column where the cold gas is cooled by using either classical Joule–Thomson valve or a turbo expander: as its pressure drops, its temperature goes below –100°C, lower than the boiling point of ethane, and then gets liquefied (all of it but commonly at least 90–95%). There is also a less efficient absorption process that can recover up to 40% of ethane and 90% of propane as the gas flows upward through a tower filled with absorbing oil; the enriched oil is then led to a distillation column where the NGLs are boiled off and oil is then recycled for repeated use.

The United States has now about 600 gas processing plant, and many central facilities have been built to treat large volumes of raw gas, both near major field and close to major market centers. For example, St. Fergus plant in Scotland processes about 20% of all natural gas from the North Sea (Total, 2014), and Aux Sable processing plant in Channahon near Chicago removes NGLs from natural gas imported from Alberta where the gas is treated to remove water and acid compounds (Figure 3.7). Importance, cost, and economic impact of natural gas treatment vary depending on the composition of extracted gas. Little has to be done with the gas from the West Sole field in the North Sea as it is a dry gas composed of nearly 95% of $CH_4$ and just 3.1% of $C_2H_6$, less than 1% of heavier alkanes, 1.1% $N_2$, and 0.5% $CO_2$ and contains no sulfur. In contrast, Qatari gas has less than 77% $CH_4$, nearly 13% of $C_2H_6$, almost 3% of heavier alkanes, and 1% of $H_2S$.

Figure 3.7   Natural gas processing plant, Central Alberta, Canada. © Corbis.

## 3.2 PIPELINES AND STORAGES

Pipelines are expensive and often surprisingly long-lived structures, usually out of sight and operating with exceptionally high reliability—but their safety requires dedicated management of corrosion risks, and their failures can have locally a truly catastrophic impact (AGA [American Gas Association], 2006). Crude oil, oil product, and natural gas pipelines share many commonalities, starting with construction methods and ending with the need for constant monitoring, but they move two very different substances. Crude oil and refined oil products are virtually noncompressible, while natural gas in its ambient, gaseous state can be compressed. Gas moving in pipelines is pressurized to between roughly 1.4 and 10.4 MPa (200–1,500 lbs/in$^2$), that is, at least about 14 times the ambient atmospheric pressure at sea level (101.325 kPa). This means that natural gas pressurized to 1 MPa will have density roughly 10 times higher than at ambient pressure (8.4 vs. 0.85 kg/m$^3$).

And while oil pipelines are a very important component of global oil industry, gas pipelines are simply indispensable. While the modern oil industry (extraction and transportation of crude oil and production and distribution of refined oil products) is highly dependent on pipelines, it is not only possible but often more economical to move large quantities of liquids by other means. The cheapest way to move crude oil between continents is in double-hulled giant tankers, and both intra- and international deliveries can use trucks, railroads, or river barges. Since 2008, the railroad option has actually expanded in North America as construction of new pipelines has not kept pace with rapid increases of shale oil extraction in North Dakota. In contrast, rail, road, barge, and tanker transport cannot be used to transport natural gas at ambient pressure and temperature. Without large-diameter long-distance pipelines, there would be no economical long-distance transmission of natural gas, and the fuel could be used only by industries located near its sources or by households, institutions, and enterprises in nearby cities.

Another key difference concerns pipeline function. Oil fields have often extensive networks of gathering pipelines; large-diameter trunk lines move crude oil to refineries, to ports, or directly to foreign buyers; and oil product lines carry gasoline, kerosene, and diesel oil or fuel oil to local storages and distributors. In contrast, small-diameter (0.5–6 in. or 1.22–15.24 cm) distribution lines bringing gas to individual consumers are the most extensive segment of the natural gas transportation system: in the United States, their length is now about 3 million km (1.8 million km of distribution mains leading to industrial consumers and 1.2 million km of service lines supplying businesses, apartments, and households). In 2012, the total length of large-diameter (4–48 in. or 10.16–121.2 cm) onshore natural gas trunk lines was 477,500 km, and field gathering lines added up to less than 17,000 km.

By the 1870s, when modern hydrocarbon extraction became a large commercial enterprise, there were extensive networks of distribution lines supplying manufactured gas to businesses and homes in most major cities, but the absence of long-distance pipelines restricted the use of natural gas not only during the last three decades of the nineteenth century but also during the earliest decades of the twentieth century. Wooden lines that were built until the 1870s by using short pieces of hollowed-out pine trunks leaked and could not withstand much pressure. The earliest (low-pressure) pipelines distributing coal gas were made of cast iron, as were the first lines bringing gas to cities from nearby reservoirs, such as the 25 km line from the Maurice River area delivering gas for street lighting in Trois-Rivières in Quebec in 1853.

Foundations for mass-scale pipeline construction were set by the widespread adoption of inexpensive steelmaking (Bessemer) process during the 1860s (Smil, 2004) and by the invention of seamless steel pipes by Reinhard and Max Mannesmann at their father's metal factory in Remscheid (Koch, 1965). In 1885, they introduced the pierce rolling process, and a few years later came pilger rolling, a concurrent lengthening of a pipe and reduction of its diameter and wall thickness. More than a century later, the Mannesmann process still dominates pipe production, but the company lost its independence; in 1999, it was taken over by Vodafone, and in the year 2000, Mannesmannröhren-Werke became a part of the Salzgitter Group (Salzgitter Mannesmann, 2014).

In 1872, a wrought-iron line of more than 80 km brought gas to Titusville, PA, and by 1886, Pittsburgh was receiving natural gas from nearby fields through wrought- and cast-iron pipes with diameters of up to 60 cm. But long-distance natural gas pipelines remained uncommon. The earliest one of these was a nearly 200 km long line from central Indiana to Chicago, completed in 1891, and the first gas export line (20 cm diameter) from Essex County, Ontario, to Detroit, crossing the Detroit River from Windsor in 1895, was less than 40 km long. Between 1906 (the first year for which the nationwide statistics are available for the US gas extraction) and the end of WWI, American gas production grew by about 85%, while much, or most, of the associated gas continued to be vented or flared. In 1920, the ratio of wasted/used gas in the United States was 0.83, but in 1930 it rose to 1.47.

Discoveries of natural gas in Texas, Arkansas, Oklahoma, and other Great Plains states during the 1920s (most notably of the giant Hugoton gas field in 1922) and the need to connect them with large Midwestern cities stimulated such pipelining innovations as electric resistance welding, electric flash welding of seams (greatly improving seam strength), and large-diameter (up to 60 cm) seamless pipes made in 12 m lengths, double the previous standard. The country's first long-distance (nearly 1,600 km, 60 cm diameter) pipeline between Texas and Chicago was completed after just 1 year of construction in 1931, followed by a line from the Texas Panhandle to Michigan (Castaneda, 2004).

## 3.2.1   Modern Pipelines

The first post-1945 gas deliveries from Texas to the Northeast used the two famous oil pipelines, Big Inch and Little Inch (both just over 2000 km long), that were built in 1942 and 1943 in order to bring crude oil and refined products to Philadelphia and New York (TETCO, 2000). The lines were decommissioned in November 1945; in 1947, they were bought from the US government by the Texas Eastern Transmission Corporation and converted to carry natural gas, first under low production pressure and later with a nearly 10-fold capacity under compression. In 1953, the Gulf gas reached New England. Concatenation of technical advances—in metallurgy, pipe making, trenching, welding, pipe protection, pipe laying and better compressors—came together after WWII to begin a new era of long-distance, large-diameter, high-throughput natural gas pipelines. The first postwar decade saw the beginning of a generation of rapid pipe laying made possible by fully mechanized trenching, by the adoption of double submerged arc processes for making longitudinal seams in 1948, and by the introduction of high-strength pipes.

Pipeline longevity was enhanced by standard coating (with coal tar or asphalt enamel, done during the installation) and by cathodic (galvanic) protection against corrosion, either passively by using sacrificial anodes or by more effective impressed current systems with anodes connected to a direct current source. These measures were largely or totally absent before WWII, but when properly applied, cathodic protection (especially when incorporating deep well ground anodes) can keep pipe's original wall thickness and strength almost indefinitely (AGA, 2006). Pipeline safety was further enhanced by the introduction of radiographic inspection of welds and (starting in the 1960s) by using smart "pigs" (instead of simple low-resolution caliper deformation tools) for internal pipe inspection. The latest pigs are equipped with high-resolution magnetic flux sensors and ultrasonic tools that make it possible to locate precisely any pipe anomalies and hence limit the required dig area.

Brush-bearing devices were used to clean pipelines for many years, and the first monitoring tools were smart pigs equipped with electronic sensors to check the thickness and integrity of pipe walls. And the 1960s also saw the introduction of plastic pipe coatings before installation, and the now standard fusion bond epoxy coatings can withstand construction activities and have a very long durability (AGA, 2006). New liquid coatings have made it much easier to recoat, repair, and rehabilitate aging pipelines and extend their lifespan at costs that are substantially lower than pipe replacement (Alliston, Banach, and Dzatko, 2002). Perhaps the most remarkable advance in construction has been a more common use of trenchless installations using guided directional drilling to go under rivers, roads, and built-up surfaces.

Advances were also made in pushing the gas through the lines. Periodic compression of transported gas is necessary in order to compensate for friction and elevation differences that would eventually slow the speed of the moving gas (between 5 and 12 m/s, which is commonly about 30 km/h) and reduce its pressure. Depending on terrain and volumes of the gas, compressor stations are spaced every 60–110 km (Figure 3.8). The world's longest export pipeline that carries the West Siberian natural gas to Europe has 41 compressor stations along its nearly 4500 km route, one every 110 km. Unless closed down by accidents, for periodic testing of the station's emergency shutdown system or for a major overhaul, these stations operate year-round. They rely on separators and scrubbers to remove any solid or liquids, and as the repressurization warms the gas, it may have to be cooled before reentering the pipeline. Exhaust silencers are used to reduce the noise generated by compressors, and backup generators are available to supply electricity in emergency.

The first prime movers installed at American compressor stations during the 1930s were spark engines with horizontal cylinders, and nearly all of them were later (between 1950 and 1970) replaced by integral units, that is, by reciprocating engines sharing the same crankshaft with compressors. These large units, expensive to operate but highly efficient, are still common at the US compressor stations, but gas turbines (both aeroderivative and industrial stationary designs with capacities of up to 15 MW) have been the best choice for all major pipelines since the 1960s (for more on these reliable machines, see Chapter 4). Introduction of reliable and efficient gas turbines has largely eliminated the need for

Figure 3.8   US compressor stations.

electricity or liquid fuels (previously the dominant choice to power compressors) as these compact machines could be powered simply by divert-
ing a small volume of the transported gas. Compressor stations on
long-distance pipelines consume typically 2–3% of the gas pushed
through a line (in comparison, resistance losses for high-voltage electricity
transmission are 6–7%).

Environmental impact of natural gas pipelines is limited. Their mechanized construction requires access by heavy trenching and pipe-laying
machinery and claims 15–30 m wide corridors. After the lines are buried
(at least 1.5 m below the surface), right-of-way strips must be maintained
to allow access for possible repairs. These strips range from 10 to 30 m,
but the land can be grazed by cattle or use for annual crops, and Canada
even allows planting of trees less than 1.8 m tall as long as they are at
least 1 m away from the line. Permafrost precludes the burial of pipelines
in the Arctic, and the lines must be placed on well-anchored steel support
and equipped with expansion joints in order to compensate large
seasonal temperature differences.

The two main factors determining the transportation costs are the
pipeline's capacity and length. For example, doubling the distance from
2,000 to 4,000 km will roughly double the operating cost per unit of
delivered energy, while doubling the annual capacity from 5 to 10 Gm³
will cut the cost by about 30%, and quadrupling it to 20 Gm³ will cut

the shipping cost in half (Messner and Babies, 2012). Where there is a choice between a pipeline and LNG, delivery pipelines are less costly for distances of up to 3,000 km for 60–70 cm diameter lines and up to 5,000 km for 140 cm diameter lines.

Given the longevity of well-built and well-maintained pipelines, it is not surprising that so many large US natural gas transmission lines date to the period of rapid post-WWII expansion, to the 1950s and 1960s. In 2012, the United States had 477,500 km of onshore natural gas transmission lines, of which 48% were built between 1950 and 1970 (with the aggregate length roughly split between the two decades) and 35% were older than 50 years (PHMSA [Pipeline & Hazardous Materials Safety Administration], 2014a). Safety of older pipelines is an obvious concern, but analysis of oil pipeline incidents by Kiefner and Trench (2001) showed that the number of years a pipeline has been in service is an unreliable indicator of its condition: what matters more are the techniques and methods used in a pipeline's construction. That is why Kiefner and Rosenfeld (2012, 30) concluded that "a well-maintained and periodically assessed pipeline can safely transport natural gas indefinitely because the time-dependent degradation threats can be neutralized with timely integrity assessments followed by appropriate repair responses."

Leaks in trunk lines and in distribution networks should be always limited to less than 1% of the transported volume: pipeline losses do not only increase the cost of delivery but are also source of a powerful greenhouse gas (for details, see Chapter 7). Unfortunately, we cannot make safety comparisons for natural gas pipelines with other modes of land transportation: pipeline transport of oil is roughly 40 times safer than in rail tanks and 100 times safer than by road tankers, but, obviously, such comparisons do not really apply to natural gas—LPG shipped by railroads in tank cars are not a directly comparable fuel category, and although both compressed natural gas and LNG can be distributed by trucks, there is no large-scale transport of these fuels in any modern economy.

Comprehensive US statistics allow us to trace the frequency of all significant and serious pipeline incidents as well as the ensuing injuries, fatalities, and property damage. The number of serious incidents (those involving a fatality or hospitalization) for transmission pipelines has declined between 1994 and 2013, and both fatalities (averaging 2/year during 20 years) and injuries (averaging 10/year) have remained low with only 2 years (2000 and 2010) having accidents with, respectively, 15 and 10 deaths (PHMSA, 2014b). Serious incidents affecting distribution lines are, as expected, more common, often caused by severed line during construction or by house fires: the total number of incidents has been about six times higher than for transmission lines, and 20-year averages were 53 injuries and 14 fatalities a year but with all of these indicators showing clearly declining trends.

The long-term annual average of all fatalities (16 a year) translates into accidental mortality of 0.005/100,000, and putting that risks in perspectives is best done by comparing it with such common fatalities and injuries as falls and car accidents: in 2012, falls in the United States caused 13.4 million consulted injuries, and transportation accidents injured 3.6 million people (CDC [Centers for Disease Control], 2014), and in 2010, 26,000 died as a result of falls (8.4/100,000), and car accidents killed 33,500, with a mortality rate of 10.9/100,000 (Murphy, Xu, and Kochanek, 2013; CDC, 2014). This means that it is roughly 2,000 times more likely to die as a result of fall or car accident than as a result of an incident involving a natural gas transmission or distribution line.

To complete the description of the entire US natural gas transportation system, I should also note that in 2012 the country had 17,000 km of onshore gathering lines, as well as 7,800 km of transmission and 9,800 km of gathering lines offshore. The entire network is made up of more than 200 pipeline systems with 5,000 reception points, 1,400 interconnection points, and more than 1,100 delivery points and is connected to national natural gas transmission systems of Canada and Mexico (Figure 3.9). There are also nearly 40 market centers (most of them in place since the 1990s) along major lines in the United States and Canada that provide gas shippers and marketers with access (both receipt and delivery) to two or more pipeline systems, as well as transportation between the centers and requisite administrative services (Tobin, 2003).

The United States also has nearly 19 natural gas multiline hubs: 12 in Texas and Louisiana, two in Illinois, and one each in California, New Mexico, Colorado, Wyoming, and Pennsylvania. Henry Hub, named after Henry hamlet near Erath in southern Louisiana (and owned by Sabine Pipe Line, Chevron's subsidiary), sits at the interconnection of nine major interstate pipelines, and it has been the centralized point for natural gas futures trading in the United States (Sabine Pipe Line, 2014). Henry Hub natural gas prices are expressed in US$ per million Btu (abbreviated as MMBtu; as already noted, unlike in the International System of Units where M stand for million, in the US usage, M is 1000 and million is MM). Henry Hub has been the primary standard for the US natural gas spot (wholesale) price, and before 2009, it had seen recurrent short-lived sudden peaks: in December 2000 to $8.9, in October 2005 to $13.42, and in June 2008 to $12.69 (USEIA, 2014c). Since

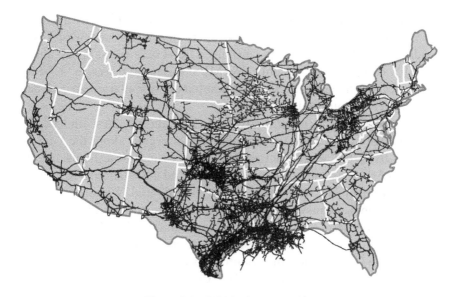

**Figure 3.9**   US pipeline network.

March 2010, their monthly mean stayed below $5 except for unusually cold February of 2014 (monthly mean $6.00, day peak of $8.15 on February 10, 2014), with the bottom at $1.95 in April 2012.

During the late 1990s, the Henry Hub price was actually slightly more expensive than the EU mean price and only about 15–25% cheaper than Japan's LNG imports. Until mid-2008, wholesale natural gas prices in the United States were only slightly lower than in Europe and about 20% lower than in Japan (EC [European Commission], 2014). The subsequent fall of American wholesale price, driven primarily by rapid shale gas extraction, led to a great decoupling. In spring 2010, the British spot price (which, too, fell since late 2008) was still similar to the declining US rate, but German spot price was double the US rate, and the Japanese LNG imports averaged nearly three times as much. These differences had widened by 2013 when Henry Hub price averaged $3.71/MMBtu, UK (Heren NBP index) price was $10.63, German imports were $10.72 (nearly three times the Henry Hub price), and Japanese LNG imports cost $16.17/MMBtu or 4.35 times the Henry Hub price (BP [British Petroleum], 2014a).

Decoupling of historically closely related crude oil and natural gas prices also took off in 2008: between 1991 and 2008, oil-to-natural gas price ratio (dollars per unit of gross energy content) fluctuated between one and two, but the subsequent rapid rise brought it briefly to more than eight during 2012. When a comparison is done with price averages for 2013—natural gas at \$3.73/MMBtu at Henry Hub and West Texas crude oil at \$97.98/barrel at Cushing, OK—crude oil supplies about 60 MJ/\$ compared to roughly 280 MJ/\$ for natural gas, a 4.7-fold difference that makes natural gas a great bargain for all uses where it can displace refined oil products.

Maps of American natural gas lines show their highest (indeed the world's highest) concentration in the coastal region along the Gulf of Mexico from the southernmost Texas the easternmost Louisiana; in southern Arkansas and northern Louisiana; and in Oklahoma. Relatively dense networks are in Kansas, Iowa, Illinois and Wisconsin in the Midwest, Ohio, Pennsylvania and upstate New York in the East, and Colorado, southern Arizona and California in the West. Columbia Gas Transmission Company (serving the Northeast) has the highest annual capacity (86 Gm$^3$) among the country's large interstate natural gas systems, but the Northern Natural Gas Company (78 Gm$^3$, with service area extending from Texas to Illinois) has longer main lines (25,400 vs. 16,600 km). Texas Eastern Transmission (66 Gm$^3$, 14,700 km of main lines) comes third, bringing the Gulf of Mexico gas to the Northeast. I will describe the world's longest natural gas pipelines in Europe and Asia in Chapter 5 when dealing with the emergence of large-scale international gas trade.

## 3.2.2   Storing Natural Gas

Storage of natural gas is an essential component of the production and distribution system. As the seasonal demand for natural gas rises, the only way to meet it by direct pipeline deliveries would be to raise their daily rates to levels that would be multiples of low-season demand, obviously an uneconomical (and also impractical) solution. And even in warm climates (where gas is not used for seasonal heating), it is necessary to cope with daily (weekdays–weekend) and intraday fluctuations caused by household demand for cooking and, much more so, by the requirements of industrial consumers. Storages required for these fluctuations are tiny fractions of volumes needed to assure adequate gas supply during winter months in more northerly latitudes (Canada, parts of the United States and the EU) where seasonal residential and commercial heating may account for as much as 90% of a city's annual gas consumption.

With rising shares of electricity generated by natural gas (see the next chapter), the demand for highly responsive storage has been increasing in all areas of all affluent countries where more natural gas is used to produce power during the peak consumption hours (often spiking to record levels due to air conditioning demand during prolonged hot spells), a practice that entails large, and often relatively rapid, fluctuations of gas supply. And adequate storages must be also in place in order to ensure uninterrupted supply in the case of upstream pipeline accidents or other breakdowns that would cut off the incoming gas for hours or days. And storage in the United States, the world's largest gas producer with fluctuating prices, is also done for profit: banking the gas at times of low prices and selling it once demand picks up.

Natural gas is not a perishable commodity, but storing it, and being able to release it on demand and in variable volumes, requires a number of specific conditions. Storing gas in a depleted oil and gas reservoir is the most obvious, hence the oldest, option used for the first time near Weland (not far from Niagara Falls in Ontario) in 1915, and it is still the most common choice. Once the gas was extracted, porous underground formations remain intact, their extent and properties have become well known over the years or decades of exploitation, and the infrastructure needed to operate the storage (wells, pipes) is already in place, a combination that makes the old reservoirs (as long as they are near major consumption centers) also the least expensive natural gas storages to set up and to operate. While the porosity of a depleted reservoir limits the volume of gas that can be stored, its permeability can be increased by using shaped charges and controlled explosions in order to raise the speed of gas injection and withdrawal.

Reservoir pressure must be maintained by keeping sufficient volume of cushion (base) gas (typically about 50% of the entire storage). That gas remains unrecoverable and the capacity of reservoirs is measured only in terms of working gas that can be withdrawn according to demand. Large volume of gas that can be held in depleted gas reservoirs makes them perfect for securing seasonal requirements, with injections done during summers and gradual withdrawals (these storages cannot deliver sudden, rapid outflows) continuing between November and March. This means that these reservoirs have usually only once-a-year turnover unlike salt caverns, the storages used to supply peak demand whose turnover is only weeks or even just days. The US now has about 350 storages in depleted reservoirs, mostly in seasonally cold Northeast and Midwest.

Salt caverns are the second best option for storing large volumes of gas: their impermeability and structural strength guarantee their integrity and longevity, but they must be first formed by drilling wells and injecting water to dissolve enough salt to create suitably sized spaces inside natural salt domes at depths mostly between 500 and 1,500 m. This is costly, but it creates a storage that has the highest reliability: as long as a cavern contains enough cushion gas (about a third of a cavern's total capacity) to keep the entire volume under sufficient pressure, the needed gas can begin to flow rapidly (in less than an hour) and at a fairly high rate, and hence, they are best used to cover peak load needs; that is why their numbers have been steadily growing along with the expansion of gas-fueled electricity generation (see the next chapter). At the same time, gas volumes stored in salt caverns are too small compared to those held by depleted reservoirs and are not used for base load storage. The United States now has about 50 salt-cavern storages, with the largest number in Texas and Louisiana, the two states with numerous salt domes.

In the regions that are far from any existing or former gas-producing fields and that are also devoid of any natural salt formation, gas must be stored in aquifers. These water-holding strata are obviously porous and permeable, but it is not easy to ascertain their overall size and porosity and to find their potential storage volume. Aquifers chosen for gas storage must have their water-bearing sedimentary formation overlaid with an impermeable rock cap in order to prevent gas escape into aquifers tapped for drinking water. Aquifer storage also requires much larger volumes of cushion gas (up to 80% of the total capacity) in order to maintain desirable rate of outflow; in some cases, high-pressure injection must be used to displace water by gas; in virtually all cases, the withdrawn gas must be dehydrated; and there may be considerable long-term losses as injected gas seeps out of the aquifer layers.

US data offer weekly or monthly monitoring of all key storage variables including capacity, storage by type, base, and working gas inventories (USEIA, 2014d). In 2012, depleted reservoirs accounted for about 85% of total storage volume, aquifers for 10%, and salt caverns for the rest; the analogical shares of working volume were about 81, 9, and 10%; and in terms of actual delivered gas, the shares were 75:10:15 (USEIA, 2014d). The total storage capacity in the United States reached to more than 240 Gm³ by 2013, equal to roughly a third of annual demand in that year, and about half of it was the working storage. Since the year 2000, actual stored volume fluctuated between October maxima of about 200 Gm³ and summer minima of about 135 Gm³.

Gas liquefaction made it possible to set up aboveground storages of natural gas: LNG tanks can store only limited volumes of the gas, but it can be instantly available. In the United States, first such facility was built in Cleveland in 1941 (for more, see Chapter 5), and by the year 2000, the United States had about 100 LNG storages (with or without liquefaction), concentrated in regions lacking large underground storages (New England and coastal Mid-Atlantic) and used to cover peak demand. In 2012, total additions to these LNG storages amounted to less than $1\,Gm^3$, a tiny fraction of underground storages but an important option for peaking demand.

## 3.3    CHANGING PRODUCTION

For decades, associated gas accompanying crude oil production was an unwelcome and unwanted by-product. Before the advent of long-distance pipelines, it could have only limited local uses: in the United States, this began to change before WWII, but in the Middle East this was true even in the early 1970s. Faisal al-Suwaidi, the former CEO of Qatargas, recalls how in 1971 the discovery of the world's largest natural gas field, the North Dome, was met with dismay: why Qatar has not been so lucky as the neighboring oil-rich Saudis? At that time, with commercial gas liquefaction in its earliest stages and with no large global market for LNG, gas associated with the extraction of crude oil could be used locally for enhanced oil recovery or burned for heat or electricity generation, but huge volumes of nonassociated gas were more of a problem than an opportunity (in the fifth chapter, I will note how al-Suwaidi engineered the gas-based transformation that made Qatar the richest state in the world).

In the absence of major local demand, the extracted gas had to be simply vented or flared, which is piped into tall stacks and burned. This wasteful and environmentally damaging practice (see Chapter 7) has been greatly reduced since the 1970s, but there are still considerable volumes of stranded gas, and the annual volume of global flaring is still unacceptably high. The first detailed global account of natural gas flaring ended up with $165\,Gm^3$ in 1970, with Iran (about $30\,Gm^3$), Venezuela (about $18\,Gm^3$), and the United States (nearly $14\,Gm^3$, and just ahead of Saudi Arabia and the USSR) wasting most of their gas (Rotty, 1974). Despite of the industry's rapid post-1970 growth, the annual rate of flaring is now less than four decades ago: in 2010, it was estimated at $134\,Gm^3$, mostly in giant Western Siberian oil fields, in Nigeria, and in Iran, but that is still an equivalent of almost 20% of the US gas use (Roland, 2010; GGFR [Global Gas Flaring Reduction], 2013). Flaring in the United States was reduced by 75% during the 1970s and then fluctuated at a new low level, but it has nearly doubled between 2000 and 2012 due to the rapid expansion of shale gas production (CDIAC, 2014; Figure 3.10).

And it is not only due to flaring that in almost all cases the volume of the marketed natural gas is substantially lower than the gross extraction (labeled gross withdrawal in the US statistics). In recent years in the United States, the average difference was almost exactly 20%, with (as already noted) venting and flaring wasting less than 1% and about 11% used to repressurize wells; 3% were nonhydrocarbon gases separated from the alkanes, and about 4% were losses during the production (USEIA, 2013a). As already explained, further (relatively small and variable) losses take place during the long-distance transportation of natural gas and its distribution, and gas actually delivered to consumers may be no more than 75% of initial withdrawals even in countries that have virtually eliminated flaring.

Global production of natural gas began to matter only after WWII. The total volume was less than $4\,Gm^3$ in 1890 and only about $7\,Gm^3$ in 1900, nearly all of it in the United States and Russia (UNDESA [United Nation Department of Economic and Social Affairs], 2013). By 1930, the total rose to $56\,Gm^3$, and in 1934, when the World Power Conference published its first Statistical Yearbook, the main producer of natural gas

Figure 3.10    Gas flaring in Pennsylvania. © Corbis.

outside North America was Russia (about $1.6\,Gm^3$), and small volumes were extracted in Austria, Poland, Argentina, Brunei, Sarawak, and the Dutch East Indies (WPC, 1934). By 1945, the global output rose to $120\,Gm^3$ (largely due to the 50% wartime increase of US consumption), and then it nearly doubled in just 5 years to $220\,Gm^3$ and doubled by 1960 to $440\,Gm^3$.

Then a more rapid and fairly linear increase brought it to $1.2\,Tm^3$ by 1975 and to just over $2.4\,Tm^3$ by the year 2000 (Smil, 2010a; BP, 2014a). Annual global production thus increased roughly 340-fold during the twentieth century, and I have integrated the best available data to come up with roughly $65\,Tm^3$ of cumulative gas production between 1900 and 2000 (Smil, 2010a). In energy terms, that was about $2.3\,ZJ$ (zeta $= 10^{21}\,J$) compared to almost $5.4\,ZJ$ for coal and $4\,ZJ$ for crude oil, which means that natural gas contributed about 20% of all fossil energies converted by high-energy civilization during the twentieth century, compared to 46% supplied by coal and 34% coming from refined liquid fuels. Cumulative totals for the second half of the twentieth century show crude oil at 39%, coal close behind with 38%, and natural gas at 23%.

During the first dozen years of the twenty-first century, natural gas extraction rose by just over 40% to nearly $3.4\,Tm^3$, and in 2012, gas supplied nearly 28% of all fossil energies and 24% of all primary commercial energy, exclusive of traditional biomass fuels (BP, 2014a; Figure 3.11). This expansion of absolute output has been accompanied by significant diversification of supply. Pre-WWII extraction of natural gas was an overwhelmingly American affair: in 1900, the US production accounted for all but a tiny fraction (<2%) of the global output; by 1950, the share was 75%; and it was still 60% by 1970. Then came the rapid rise of Soviet gas extraction (it had more than quadrupled in two decades between 1970 and 1990), European output (mainly Groningen and the North Sea gas from Norwegian and British waters) doubled during the 1970s before it stabilized, and the Saudi production had tripled during the 1990s. This brought the US share of the global output down to 25% by 1990.

Afterward, the output in the countries of the former USSR fell and stagnated before it began to recover after the year 2000, but a number of old producer multiplied their extraction: between 1990 and 2010, the absolute increases were 2.2-fold in Australia, nearly four-fold in Malaysia, more than six-fold in China and Iran, nine-fold in Trinidad and Tobago, and 18-fold in Qatar, and by 2010, the US share of the global natural gas production was just below 19%. Another way to illustrate the diversification process is to trace the number of major

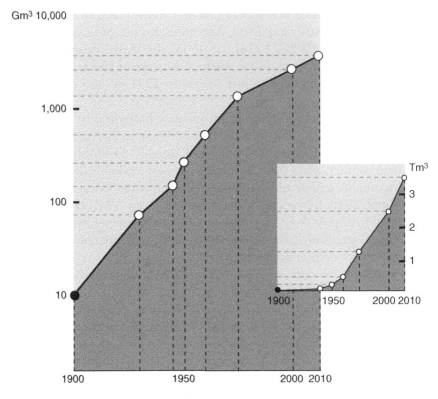

**Figure 3.11**   Global natural gas extraction.

(annual output >20 Gm³) natural gas producers: only five countries were in that category in 1970 (the United States, Canada, the Netherlands, Romania, and the USSR), eight in 1980 (adding Mexico, the United Kingdom, and Norway), 18 in 1990 (with Russia, Ukraine, Uzbekistan, and Turkmenistan dividing the previous Soviet total), 24 in the year 2000, and 27 in 2012.

In that year, more than 50 nations (i.e., every fourth country worldwide) had nonnegligible natural gas industry (with annual output >1 Gm³), and this number will grow further due to recent discoveries. New offshore discoveries have already put Mozambique's gas reserves on par with those of Indonesia, while new fields in the Mediterranean (Tamar, discovered in 2009, and Leviathan, discovered in 2010) have made Israel a new gas exporter: in 2014, it signed deals with the Palestinian Authority and with Jordan. Moreover, for about half of the world's two dozen major hydrocarbon producers, natural gas is now more important (in total energy terms) than crude oil. In 2012, energy content of the two fossil fuels was nearly equal in Russia and Algeria, but gas was more important than oil in Argentina, Australia, Egypt, India, Indonesia, Malaysia, Netherlands, Norway, Qatar, Turkmenistan, and Uzbekistan.

This is an apposite place to note the changes in energy return on investment (EROI). This ratio—energy returned from one unit of energy invested in an energy-producing activity—has been promoted by some of the practitioners of ecological economics, but it has been ignored by most energy economists (Hall, 2011). As a result, we have only a relatively small number of EROI values for natural gas. Guilford, Hall, and Cleveland (2011) estimated EROI for the US oil and gas production during the twentieth century: the ratio shows a slowly declining trend from about 20:1 in 1919 to 8:1 in 1982 (the year of the peak drilling), followed by a recovery to 17:1 between 1986 and 2002, but in the first decade of the twenty-first century, the mean ratio fell rapidly to just 11:1.

And Grandell, Hall, and Höök (2011) estimated that EROI of Norwegian oil and gas production rose from 44:1 in the early 1990s to a maximum of 59:1 in 1996 and then it declined to about 40:1 by 2008. But as these studies combine the returns for two distinct fuels, their EROI ratios can be used only as indicators of a basic (and expected) secular trend of diminishing, but still relatively high, returns. But we have a specific study of Canadian natural gas production during the last decade of the twentieth century and the first decade of the twenty-first century: according to Freise (2011), EROI in Canada's conventional natural gas has entered the era of permanent decline after falling from 38:1 in 1993 to 15:1 in 2005.

# 4

# Natural Gas as Fuel and Feedstock

Natural gas was not the first gaseous substance with widespread commercial use: its combustion by household, industries, and commercial establishments was preceded for nearly a century by the burning of manufactured gas. This fuel was first generated during the latter half of the eighteenth century as a by-product of coking. The first dedicated facilities producing the gas for lighting (by carbonization of bituminous coal, i.e., by high-temperature heating of the fuel in ovens with limited oxygen supply) began to deliver gas for industries, workshops, and street illuminations in 1812 in London, 1816 in Baltimore, 1818 in Brussels, 1825 in New York, 1829 in Boston, and 1849 in Chicago (Webber, 1918). By the middle of the nineteenth century, every major city in Europe and the United States had its gas works, and in 1855, the introduction of Robert Bunsen's burner (mixing the gas and air in correct proportion to produce safe flame for cooking and heating) opened the way for mass-scale domestic adoption of town gas, and between 1860 and 1900, the production and distribution of town gas became established as one of the leading industries of new industrial era (Hughes, 1871; Figure 4.1).

Ammoniacal liquor and tar liquid fraction had to be separated from the generated coal gas—actually a mixture of gases dominated by hydrogen (48–52%), with 28–34% of $CH_4$, 13–20% of $CO_2$, and 3% of CO with energy density no higher than $20\,MJ/m^3$ (<60% of typical natural gas)—whose combustion provided a very inefficient source of

*Natural Gas: Fuel for the 21st Century*, First Edition. Vaclav Smil.
© 2015 John Wiley & Sons, Ltd. Published 2015 by John Wiley & Sons, Ltd.

**Figure 4.1**   New York 1900: light and cook with gas. © Corbis.

exterior and interior lighting. But because the earliest incandescent lights were also very inefficient and because mass adoption of electricity required first the extension of transmission lines and interior wiring, coal gas continued to be used in many cities for decades after the first urban power plants were built in the United States and Europe during the 1880s.

In many large cities, the town gas industry was active throughout the first half of the twentieth century, and in some places, gas works operated even into the 1950s and 1960s. Town gas lost first its lighting market to light bulbs (and, starting in the 1930s, to fluorescent lights), but it still competed with electricity for household and institutional cooking. Once natural gas became available (in many US cities before WWII, in Europe widely during the 1970s), town gas era ended, but not its long-lasting environmental legacy of persistent tars that are still contaminating soils and water near former gas works, with the United States having more than 50,000 such sites (Hatheway, 2012).

Because natural gas combines three attributes that are very desirable for a fuel—clean combustion, convenient regulation of the combustion process, and (once the pipelines are in place) on-demand delivery—it has been a favorite (and often superior) form of energy for a variety of uses. Not surprisingly, the extent of these uses has changed with time, and it has been also influenced by the presence or absence of infrastructures (natural gas could be readily distributed in Western cities where houses were connected for decades to receive municipal coal gas), assured availability (as demonstrated by location of many large petrochemical plants close to major gas fields), and affordability (despite technical advances, LNG remains inherently more expensive than pipeline gas).

Traditionally, by far, the most important use of natural gas as fuel was to generate heat or steam required for a large variety of industrial processes, from heavy metallurgy to fine manufacturing. Emergence of this market during the closing decades of the nineteenth century was followed by the adoption of natural gas as a leading fuel for space heating and cooking in households as well as in many commercial and institutional facilities, and since the 1980s, the last two sectors have also used natural gas for central cooling. Another major market for gas as a fuel came with the growth of cleaner electricity generation as methane replaced coal, and also fuel oil, in large centralized power and created new possibilities for more decentralized power based on gas turbines.

A relatively small share of global natural gas extraction that is not burned but serves as a feedstock is not a good indicator of the fuel's importance as a raw material. Production of the world's most important fertilizer now depends overwhelmingly on inexpensive natural gas: there are other ways to synthesize ammonia, but none is as economical as using methane both as a feedstock and as a fuel to energize the synthesis. And, in this world suffused by plastic materials, methane and natural gas liquids (ethane, propane, butane, and pentane) are the simplest precursors of so many complex compounds that the material circumstances of modern society would be quite different if those hydrocarbons could not supply the foundation of a large part of the modern petrochemical industry.

In addition, methane is the most important starting material for the synthesis of methanol (methyl alcohol) which, in turn, is an important chemical feedstock used in the production of formaldehyde, acetic acid, and other intermediates that are eventually turned into resins, plastics, paints, adhesives, and silicones. And methanol production is, at least for now, the most important process of transforming gas to liquids, a class of conversions designed to produce more valuable fuels that are also easier to transport. Production of gasoline, kerosene, and diesel oil has been the ultimate goal, and although technically feasible it faces many engineering and economic challenges. Before looking at major consumption sectors, I will summarize the US and foreign pricing in a few paragraphs.

As expected, high-volume industrial consumers get the cheapest gas, commercial establishments pay more, and residential consumers buy the most expensive gas. In 2013 in the United States, the three rates averaged, respectively, $4.66, $8.13, and $10.33/Mcf. The 2013 ratio of these prices 1:1.74:2.21 shows that natural gas has been getting relatively more expensive for commercial and residential consumers: in the year 2000, the ratio was 1:1.48:1.75 (USEIA [US Energy Information Administration], 2014c). The value of natural gas at the point of consumption (all totals are in current prices) rose from $837 million in 1945 to $8.5 billion in 1975, and with price deregulation, it increased to $30 billion by 1990 and $70 billion in the year 2000, and in 2013, it was roughly $113 billion with $51 billion from residential, $27 billion from commercial, and $35 billion from industrial sales (USEIA, 2014c).

But American households still pay a much lower price than the Europeans or consumers in East Asia. In 2012, when the US price ($10.71/Mcf) prorated to $0.38/m$^3$, the average residential price in EU-27 was $0.83/m$^3$ (EC [European Commission], 2014). In 2013, when the US household price was $0.37/m$^3$, the European residential prices in 22 of the continent's capital cities averaged $0.95/m$^3$, ranging from $2.66 in Stockholm (all gas imported) to $0.40 in Bucharest (still nearly 90% of total consumption from domestic sources), that is, 2.6 times higher than in the United States (Energy Price Index, 2013). Japanese households have to pay even more for their imported LNG, nearly $1.7/m$^3$, roughly 4.6 times the US rate.

## 4.1  INDUSTRIAL USES, HEATING, COOLING, AND COOKING

The category of industrial gas uses is quite heterogeneous. Energy industries account for a large share in all countries with substantial hydrocarbon production: in the preceding chapter, I have already described how gas is used in field operations (including secondary recovery of crude oil), in gas processing plants, and in powering compressors to move the gas through the pipelines. A new and important use of natural gas in oil and gas industry is its critical role in producing liquid fuels from Canadian tar sands. Two major modes of extraction are excavation of tar sand in open mines and *in situ* recovery, either by cyclic steam stimulation or by steam-assisted gravity drainage: natural gas is burned to produce the needed steam and between 25 and 35 m$^3$ of gas is required to produce a barrel of bitumen (NEB [National Energy Board], 2014a).

As already noted, undoubtedly the earliest, and well-documented, use of natural gas was to evaporate brines in Sichuan, China's most populous landlocked province (Adshead, 1992). That practice dates to the Han dynasty (200 BCE), and although it continued well into the early modern era, it has not been replicated anywhere else, and in the West, natural gas was used first (during the closing decades of the nineteenth century) for street lighting and as an excellent source of heat for many industrial processes. The first application was quickly eliminated by electric lights, but industry has remained the single largest user of natural gas ever since. The earliest reliable consumption breakdown for the US natural gas consumption shows that in 1906 industrial uses claimed roughly 72% of total sales, with the rest used by households and commercial enterprises (Schurr and Netschert, 1960). In Canada, Edmonton was the first city to switch to natural gas (delivered by 130 km long pipeline from Viking, Alberta) for heating and cooking in 1923.

## 4.1.1   Industrial Uses of Natural Gas

In 1950, the US industries used about 60% of all natural gas, and by 2013, the share declined to about 34%, but it was still slightly ahead of 31% used for electricity generation (USEIA, 2014e). Moreover, natural gas consumption by the industrial sector represents a larger fraction of the final energy use than for other sectors whose energy use is heavily dominated either by electricity (more than 75% in commercial sector, more than 66% by households) or by liquid fuels (more than 95% in transportation). Review of sectoral fuel inputs in American economy showed that natural gas accounted for about 55% of the total energy requirement in the fabrication of metal products, 52% in food industry, nearly 50% in metal casting and transportation equipment, and 45% in chemical industry, with the average of about 36% for all industrial sectors reported in the *Manufacturing Energy Consumption Survey* (ICF International, 2007).

In absolute terms, the largest consumer of natural gas is chemical industry followed by petroleum refining, food manufacturing, and pulp and paper. The three most common categories of use include the production of hot water and steam required for many kinds of industrial processing; the combustion of natural gas for space heating, centralized air conditioning, waste treatment, and incineration; and the production of direct heat for preheating and melting of materials (particularly metals), for drying, and for dehumidification. Gas-fired desiccant systems assure that moisture in materials and products is lowered below the levels that could cause defects. Many industries also use natural gas to supplement combustion of other fuels: this cofiring with wood and other biomass or coal increases overall efficiency and reduces emissions. General quest for cleaner production—virtually no particulates and no $SO_x$ emissions, effective control of $NO_x$, and reduced greenhouse gas generation—means that with greater gas availability, the dependence of the industrial sector on natural gas is likely to increase.

Particularly, efficient uses of natural gas include infrared heating units and direct-contact water heaters. Standard industrial water heaters and boilers raise the temperature of water indirectly, by using heat exchangers with the liquid enclosed in vessels or tubes, and they operate with efficiencies of no more than 70%. In direct-contact water heaters, cold water is introduced at the top of a unit, and as it flows down through packing (stainless steel nodules, rings), it is heated by combustion gases rising from a burner at the bottom of the unit, and the overall efficiency can be as high as 99.7%. Direct-contact water heaters are used for space heating, in textile and food manufacturing, and also in large laundries (Quik Water, 2014).

Food production needs large volumes of heated water, steam and dry heat for cooking, baking, fermenting and sterilizing, and virtually all canned and conserved items, as well as dairy products, juices, wines and vinegar must undergo pasteurization, that is heating to 72°C to eliminate harmful pathogens. Depending on the requirements, this may be done during the processing or as the last production step before releasing a product. By far the highest demand for heat in paper manufacturing is for drying wet paper: it requires removing 1.1–1.3 kg of water for every kg of paper, and it is done by contact with steam-heated cylinders (Ghosh, 2011).

Kilns used to fire bricks, stoneware, and ceramics as well as fine china are commonly heated by natural gas. In the United States, 80% of bricks are made in gas-fired kilns where the temperature reaches up to 1,360°C during the last (vitrification) stage of the firing (BIA [Brick Industry Association], 2014). But natural gas has been usually too expensive for much more massive production of Portland cement: even in the United States, the sector has relied mostly on coal, petroleum coke, and combustible waste, with natural gas supplying only about 5% of all energy. Production of primary iron in blast furnaces also remains highly dependent on coal, specifically on coke produced by carbonization (high-temperature destructive distillation) of coal in oxygen-poor atmosphere, but during the twentieth century, the metal's smelting has seen more than a 60% decrease in coke-charging rates due to supplementary injection of pulverized coal, oil, or natural gas (de Beer, Worrell, and Blok, 1998).

Fabrication of steel parts uses natural gas for a variety of heat treatments including hardening (heating to a prescribed temperature followed by rapid cooling by quenching into oil, water, or brine, most commonly done for steels), tempering (to reduce brittleness by heating to a specified temperature and then letting the metal to cool), annealing (to make metal more ductile and to refine their grain structure by heating to a specified temperature, holding it for a desired time, and then cooling it slowly), and case hardening deposition of more carbon into steel surfaces by heating in the presence of a specific material. Nonferrous metals undergo annealing and solution heat treatment (heating followed by rapid quenching).

## 4.1.2 Natural Gas for Space Heating and Cooling

Before WWII, the North American wood stoves were still common in many rural areas, and urban houses were heated mostly by coal stoves in individual rooms or centrally by basement coal-fired boilers that heated water circulating through cast iron radiators or (after 1885) by warm air rising through ducts by natural convection from riveted-steel coal furnace introduced by Dave Lennox. Furnaces heated by fuel oil were the next option before the introduction of the first forced air coal-fired furnace in 1935. After WWII, the forced air heating became dominant, but the fuel changed rapidly from coal to natural gas.

These furnaces heat incoming fresh air and distribute warm air through ducts underneath floors and inside walls. They are equipped with a pilot light and an electric motor to force the heated air, they became common after 1950, and their evolution has been marked by a shrinking size and an improving performance. North American natural gas furnaces are rated by their annual fuel utilization efficiency (AFUE). Pre-1975 furnaces had AFUE of just 55–60%, wasting 40–45% of all purchased natural gas in hot exhaust gas escaping through a lined chimney. Improved versions of so-called mid-efficiency furnaces had eventually converted as much as 78–82% of energy in gas into household heat.

In Canada, sales of mid-efficiency furnaces ended on December 31, 2009, and only high-efficiency (condensing) furnaces (AFUE of at least

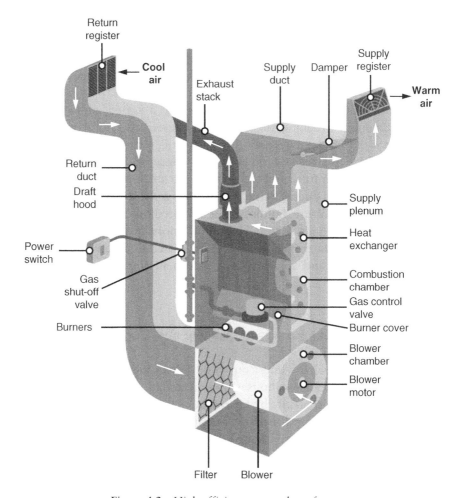

**Figure 4.2** High-efficiency natural gas furnace.

90%) and furnaces with Energy Star label (AFUE of at least 95%) are available (Figure 4.2). They are microprocessor controlled; have electronic ignition (hence no pilot light), secondary heat exchanger, and high-efficiency fan motor; and can be turned on and off by a programmed thermostat placed in a living area. Their heat loss is so small that there is no need for a chimney as warm $CO_2$-laden exhaust gas is led out through a polyvinyl chloride (PVC) or ABS plastic pipe on a side of a house and water generated by combustion is disposed into a floor drain. Many homes also use natural gas for heating water, and again, the latest heater designs have efficiencies superior to the previous models.

Post-WWII expansion of commercial (offices, stores, shopping malls) and institutional (schools, universities, hospital, museums) space created a large new category of heating preferably served by natural gas. In 2013, the US commercial sector consumed more than eight times the volume of the gas than it did in 1950, an equivalent of two-thirds of household consumption (in 1950, it was equal to less than a third). And commercial and institutional users have been also the pioneers of natural gas-powered cooling, an option that is particularly attractive for large facilities with high daytime (schools, offices) or often nearly constant cooling loads for hospitals, hotels, and supermarkets in hot climates (ESC [Energy Solutions Center], 2005; Uniongas, 2013). The choices of natural gas-powered cooling equipment include quiet and efficient absorption chillers (they can also use waste from other on-site burning of natural gas), engine-driven and steam turbine-driven chillers, and desiccant dehumidification (making high temperatures more bearable, reducing frost coating in supermarket chillers, keeping ice rinks fog-free).

### 4.1.3   Cooking with Natural Gas

Cooking (and water heating) with town gas began to make the first inroads in large Western cities during the last two decades of the nineteenth century as the town gas producers looked for replacing sales lost to electric lighting. In the United Kingdom, one out of four urban households used cooking gas by 1898 and one out of three in 1901 (Fouquet, 2008). Gas ranges became common in all major cities of the industrializing world during the first two decades of the twentieth century, and by 1930 in the United States, they outnumbered (burning town gas as well as natural gas) wood and coal stoves by nearly two to one. Cheaper electricity limited their post-WWII adoption in North America, but, even so, at the beginning of the twenty-first century, about one-third of US households cooked with gas.

Advantages of cooking with gas include more even heat distribution (flame heating also the sides of cooking vessels) and hence faster completion of a task; more precise, instant control of heat intensity (always preferred by professional chefs, particularly for rapid Chinese-style cooking); and the ability to change temperature rapidly and to cut off heat instantly (hence being able to leave pots and pans resting on the stove). But gas ranges are more expensive than the electric ones, less easy to clean (particularly when compared to new electric flat tops), not as good for baking (combustion releases water), and potentially more dangerous (risks associated with open flame, leaks, gas escaping after accidentally quenching the flame), and they can be, especially when used without an exhaust hood, a nontrivial source of indoor air pollution.

Logue et al. (2013) found that in southern California typical gas range-generated indoor exposures to CO exceed the standards for ambient air and that hoodless cooking also causes excessive $NO_2$ levels. Data collected as a part of the European Community Respiratory Health Survey showed no significant association between respiratory symptoms and gas cooking in males, but for females there, compared to electric cooking, symptoms suggestive of some airway obstruction (Jarvis et al., 1998). A more recent Norwegian study found that gas cooking produces higher levels of cancer-causing fumes than does electric cooking, but all recorded levels were below permissible occupational thresholds (Sjaastad, Jørgensen, and Svendsen, 2010).

## 4.1.4   Liquefied Petroleum Gas

Finally, a couple of paragraphs are presented on liquefied petroleum gas (LPG, mixture of propane and butane) as fuel. Many of its industrial uses are identical to the uses of natural gas as a fuel for ovens, furnaces, and kilns, in process and water heating, and high-temperature treatment of metals, but the mixture is also used in oxy–propane cutting and welding and as an aerosol propellant for a variety of household products ranging from cosmetics to insecticides. On construction sites, this portable fuel is used to heat buildings and bitumen during road repair. LPG's many agricultural applications include space heating of animal barns, grain and fruit drying, and also powering of agricultural machinery (for more on this mobile use, see Chapter 7).

LPG is also used for a range of leisure activities. The best known use is portable fuel for cooking—be it for occasional camping or as the fuel for ranges in recreational vehicles—available in a variety of refillable metallic cylinders, but there are also LPG-fueled boats (because of its minimal impact on the water if spilled, it is a better choice than gasoline), and the modern ballooning industry depends on powerful LPG-fueled burners to lift and to maneuver large balloons used for aerial sightseeing. And while there are no LPG-fueled cars in North America, there are now 4.5 million LPG vehicles in Europe (AEGPL, 2014). Again, I will look in some detail at this mobile application in Chapter 7.

## 4.2   ELECTRICITY GENERATION

Once natural gas became available at affordable prices and in larger volumes, electricity generation was its next obvious market, but long-established coal- and oil-fired generation (and, starting in the 1960s, expansion of nuclear generation) has made it a slow process. In 1950, even in the United States, after more than half a century of expanding natural gas extraction, twice as much natural gas was destined for house-holds than for electricity generation, the residential use was still 55% higher in 1975, and combustion of natural gas for electricity generation surpassed the domestic uses for heating and cooking only during the last year of the twentieth century. Until the late 1950s, virtually all gas-fired electricity-generating capacity was in medium- and large-sized central power plants where gas was burned in boilers to produce steam for turbogenerators.

Between 1950 and 2000, consumption of natural gas for electricity generation rose more than eight times, and between 2000 and 2012, it rose by 75%. In 1950, natural gas contributed almost exactly 20% of all primary fossil fuel energies used to generate American electricity (coal dominated with 66%). By 1975, its share rose only to 21%, but the total gas consumption had quintupled. Subsequent rise of gas turbines made them the preferred choice for gas combustion in electricity generation, and these highly efficient and reliable machines now dominate the market in all major economies. By the year 2000, the share dropped once again below 20%, but (starting in 2007) the increased availability of cheaper gas pushed the share to 33% in 2013. In terms of actual generation, the share originating from natural gas is lower, about 27% in 2013. And the global share is not that much lower: in 2010, 21% of the world's electricity was produced by burning natural gas, either in large boilers or in gas turbines.

The difference between the shares of installed capacity and actual gas-fired generation is mainly due to the rising importance of gas turbines. As I will show in the following pages, turbines burning natural gas have unequaled efficiencies (when working in a combined cycle with steam turbines up to 50% higher than steam turbogenerators in large coal-fired stations)—but their capacity (load) factor is relatively low because they cover mostly peak and emergency demand unlike the steam tur-bines in large central power plants steam turbines that cover the base load. Calculations based on the US data for 2012 show average load factor of 55% for coal-fired but only 33% for natural gas-fired electricity generation (USEIA, 2013a).

Central gas-fired power plants—supplied by a pipeline from a nearby field or located next to LNG regasification terminals and converting typically 33–35% (for the older stations) and 35–38% (for post-1990 projects) of gas to electricity—remain important in several countries, most notably in Japan which has the largest group of such high-capacity (>2 GW) stations originally built to replace coal- and oil-fired plants and to improve air quality in the country's densely populated coastal conurbations. Nearly 20 of these LNG-based high-capacity stations, located on reclaimed land adjacent to receiving terminals, are easily identifiable by their tall boilers and generator halls and chimneys and gas storage tanks. Japan's largest gas-fired station is 5.04 GW Futtsu in Chiba prefecture, across the Tōkyō Bay from the capital (see Fig. 5.7). The second largest station, 4–8 GW Kawagoe, is in Mie Prefecture east of Ōsaka.

Other East Asian LNG importers with large gas-fired stations include China, Taiwan, and South Korea, while Russia, Australia, and Malaysia rely on power plants burning inexpensive domestic gas. With 5.597 GW of installed capacity, Russia's Surgut-2, in West-Central Siberia (Khanty-Mansiysk region), was the world's largest gas-fired station in 2014, while New York's 2.48 GW Ravenswood station, burning natural gas, fuel oil, and kerosene illustrates the compact nature of such large facilities as it occupies a rectangular lot of just 12 ha just south of the Roosevelt Island Bridge on the East River in Queens. But in the United States, smaller-capacity gas turbines have been dominant for decades: in 2012, gas-fired boilers producing steam for large turbogenerators accounted for less than 20% of the US gas-fired generating capacity, nearly 30% of it was in simple cycle gas turbines and 52% on combined cycle gas turbines (CCGTs), and hence in the rest of this section I will describe the rise and the performance of these remarkable machines.

## 4.2.1   Gas Turbines

The quest for combustion turbines goes back to the end of the eighteenth century, to John Barber's patent that outlined an essentially correct principle but had no chance to be transformed into a working machine. Doing that remained a challenged for generations to come. Only in 1903, Aegidius Elling's six-stage turbine with a centrifugal compressor generated a bit of net power, and between 1903 and 1906, the best prototypes of centrifugal turbines built by the *Société des Turbo-moteurs* reached a net efficiency of less than 3%, far inferior compared to the best steam engines of the day. At the same time, Brown Boveri, a pioneer of steam turbine manufacturing, worked on its first gas turbine prototype (Smil, 2006).

The first practical advances came only during WWI. GE's new turbine research department, established in 1917 by Sanford Moss (whose first gas turbine design went back to 1895), developed a turbo supercharger driven by hot exhaust gases from reciprocating Liberty engines that powered America's wartime planes. More proposals and patents for turbojet engines followed during the 1920s, but the only practical application was a small gas turbine designed by Hans Holzwarth and built by Brown Boveri for a German steel mill. Real commercial breakthroughs began only during the 1930s, helped by advances in materials and by greater market opportunities; at the same time, dedication and persistence of two young engineers, Frank Whittle in the United Kingdom and Hans Joachim Pabst von Ohain in Germany, led to the first turbines (jet engines) suitable for flight (Smil, 2010a).

The first stationary gas turbine with utility-scale capacity was a machine built in 1939 by Brown Boveri for the municipal electricity-generating station in Neuchâtel (Alstom, 2007). Its rated capacity was 15.4 MW, but because its compressor consumed almost 75% of the generated power and because all exhaust heat was vented, the actual available capacity was no higher than 4 MW, resulting in a poor efficiency of just over 17%. Remarkably, the turbine decommissioned only in 2002, after 63 years of operation. Post-WWII development of gas turbines was slow: except for the US natural gas was not readily available, utility markets everywhere were dominated by coal-fueled steam turbines or hydro-turbines, and inexpensive exports of the Middle Eastern oil offered a new fuel alternative for electricity generation. Combustion of fuel oil in large boilers became an important component of electricity in coal-deficient countries as well as in Japan whose economy began its rapid expansion.

In the United States, both Westinghouse and GE introduced their first gas turbines for electricity generation (with capacities of <1.5 MW) in 1949, and a decade later, the aggregate capacity of American gas turbines was just 240 MW, less than a single large machine delivers today. In 1960, the highest turbine rating reached 20 MW, and the total installed capacity rose to 840 MW in 1965; before that year ended, an unexpected event ushered in a new era of rapid gas turbine expansion. On Tuesday of November 9, 1965, the Northeastern United States—extending from New Jersey to New Hampshire, and also parts of Canada's Ontario, overall an area of more than 200,000 km$^2$—suffered a power blackout, with about 30 million people losing electricity for up to 13 hours (USFPC [US Federal Power Commission], 1965).

Utilities had realized that rapidly deployable gas turbines would have made a critical difference. Although a turbine has to be started by external means (a small motor will do), its output can reach full load in minutes compared to hours for steam turbine. American utilities installed 8 GW of new gas turbine capacity in just 3 years after 1965 and expanded aggregate gas turbine power more than 30-fold in a decade to nearly 45 GW in 1975. This rapid expansion ended due to the rising prices of natural gas (a consequence of OPEC's quintupling of oil prices in 1973 and 1974) and declining demand of electricity (after years of rapid expansion).

Growth had resumed after hydrocarbon prices fell in the mid-1980s, and by 1990, the US utilities installed almost half of their new capacity in gas turbines, and the orders for new gas turbines have surpassed the worldwide orders for steam turbines (Valenti, 1991). No less importantly, this quantitative growth was accompanied by impressive qualitative improvements. Gas turbines are among the most reliable modern machines: when properly maintained, stationary turbines require overhaul after no less than 25,000 hours of operation (i.e., 2.85 years of nonstop operation). This reliability makes them also a preferred choice in continuous industrial applications: oil and gas industry uses them to drive pumps and compressors in processing plants, refineries, and pipeline pumping stations.

And they are also very efficient. As already noted, in 1939, Brown Boveri's first turbine had an efficiency of only about 17%; three decades later, efficiencies came close to 30%; in 1976, GE's largest (100 MW) turbine rated about 32%; and by the year 2000, gas turbines reached thermal efficiencies of just above 40%. Plants using the best steam turbines and the best simple cycle gas turbines had thus a very similar performance, but gas turbines could do much better as a part combined cycle. The early machines had low capacity (<10 MW), and their exhaust gas had temperature below 750°C, a combination that did not allow to do more than use the waste heat for preheating air for a boiler or heating boiler feedwater.

Larger units (50 MW in 1970, 100 MW by the mid-1980s, 200 MW by the mid-1990s) and higher firing and hence higher exhaust temperatures (the former ones getting above 1,000°C by the late 1960s and reaching 1,250°C by 1990; the latter ones rising from less than 400°C in the early 1950s to as much as 600°C in the latest machines) provided an unprecedented opportunity to either raise the efficiency of electricity generation or to maximize the overall efficiency of a gas turbine-based system used for cogeneration, that is, for combined production of electricity and heat (space or processing) for industrial or household uses.

## 4.2.2   CCGTs

Predesigned combined cycle plants entered the market during the late 1960s and the early 1970s under the names Steam and Gas (STAG by GE) or *Gas und Dampf* (GUD by Siemens) and Power and Combined Efficiency (PCE by Westinghouse), and eventually, the technique became generally known as CCGT. After a relatively rapid maturation, CCGT became the most common choice for both adding new capacity and repowering older plants (Balling, Termuehlen, and Baumgartner, 2002; Rao, 2012). This logical arrangement couples a gas turbine with a steam turbine: gases leaving a gas turbine are hot enough to raise steam (in an attached heat recovery steam generator) whose expansion runs a coupled steam turbine. Efficiency of this combined cycle is calculated by summing up the specific efficiencies of the two turbines and then subtracting their product: 42% efficiency for a gas turbine and 32% efficiency for a steam turbine yield a combined efficiency of just over 60%.

That has been the state of the art, but future improvements, up to 68–69%, might be possible with further combinations (adding solid oxide fuel cells or solar photovoltaics). Maximum CCGT capacities have now reached the level of midsize (400–600 MW) steam turbogenerators in large central power plants. Siemens introduced the world's largest machine in 2007: its SGT5-8000H rated 340 MW and powered 530 MW combined cycle plant with an efficiency of 60%, and it has since increased the maximum simple cycle capacity to 375 MW in simple and 570 MW in combined cycle (Siemens, 2014). GE's two latest models, 9HA 01 and 9HA 02 (introduced, respectively, in 2011 and 2014), are the current record holders with 397 and 470 MW in simple cycle, and the larger machine helps to deliver 710 MW gross and 701 MW net in combined cycle (GE, 2014a; Figure 4.3).

**Figure 4.3**   GE gas turbine. Reproduced courtesy of General Electric Company.

Cogeneration (also known as combined heat and power (CHP)) does not convert waste heat into electricity but uses it directly for processing in adjacent industrial plants or, more often, to heat water for numerous industrial uses (particularly in food and textile industries) or for space heating, such as centralized heating plants (in the past fueled by coal or by fuel oil, now also by waste biomass). Besides higher overall efficiency, cogeneration has the advantage of reduced air pollution and enhanced local and general security of supply. CHP has been fairly common in urban areas of Europe and Japan and now is also widely used in China's rapidly growing cities. In Europe, cogeneration now produces more than 10% of all electricity, with natural gas being the dominant fuel in most countries using CHP, particularly in the Netherlands, Spain, Italy, Germany, and the United Kingdom (COGEN Europe, 2014).

A particularly interesting case of cogeneration has been its use in Dutch greenhouses that now cover more than $60 \, km^2$ of the country. The practice began in 1987, it saves up to 30% on total energy costs, and the combustion of natural gas provides not only light and heat for the country's extensive greenhouses but a part of the generated $CO_2$ that is used to enrich the enclosed atmosphere to at least 1,000 ppm (compared to the global ambient atmosphere average of 400 ppm in 2014). This practice results in faster growth and higher yields of peppers and tomatoes, the two dominant products of Dutch greenhouse farming (Wageningen UR, 2014).

A new option for natural gas-based electricity generation became available with the introduction of less massive aeroderivative turbines, jet engines modified for stationary application (Smil, 2010a; Langston, 2013). GE's first design (LM6000 with rated capacity of 40.7 MW and efficiency of 40%) was based on a long-serving CF6 engine used to power Boeing 747 and 767 as well as several Airbus models (GE, 2014a). By 2013, GE offered models rated from 16.4 to 105.8 MW in both 50 and 60 Hz series (GE Power & Water, 2013). The other two large jet engine makers, Rolls Royce and Pratt & Whitney (P&W), offer their lines of aeroderivative gas turbines—the former's line based on its widely deployed Trent series of jet engines and the latter's on JT8D engine which powered first Boeing 727 and 737 planes during the 1960s. P&W has also developed packaged gas turbine designs that come fully assembled on trailers and that can start generating electricity in less than a month after reaching their location.

P&W's FT8 27 MW gas turbine is mounted on a steel base, comes packaged with its ancillary support systems, directly drives the generator package, fits on just two trailers, and it is ready to generate only 8 hours after arriving at a site (PW Power Systems, 2014).The turbine, its control trailer, access roads, fuel and electricity connections, and a safety perimeter buffer occupy only $600\,m^2$, while a 60 MW SwiftPac on concrete foundations needs $700\,m^2$, and it generates just 3 weeks after delivery. Compactness of gas turbines makes it easy to site them within the boundaries of existing central power plants, obviating often contentious approval process for new locations. For example, siting of the original Didcot-A 2 GW coal-fired station completed in Oxfordshire in 1968 met a great deal of local opposition, but gas turbines of Didcot-B, 1.36 GW gas turbine station completed in 1997, are virtually hidden within the Didcot-A site where it takes up less than 10% of the original station's area (Smil, 2015).

Ubiquity of gas turbines is attested by their rising numbers and by their share of newly installed capacities. In 2013, there were more than 26,000 gas turbines worldwide, working as simple cycle machines, as CCGTs, and in CHP system, with North America having nearly a third of the total and Asia a fifth (Platts, 2014). In 2012, the United States had 121 GW installed in gas turbines, or about 15% of all fossil fueled and 11% of all net summer capacity, but they generated only 3% of all electricity. That very low share is easily explained by their use during the times of peak power demand: during off-peak hours, their capacity factors are usually just 1–3%, while during the peak hours (in the United States now between 1 and 5 p.m.), they go above 10%, and in the Northeast (Northeast Power Coordinating Council, where the peak is extended from 10 a.m. to 20 p.m.) and in Texas (with more pointed peak centered on 3 p.m.), they rise above 25% (USEIA, 2014f).

The peaking role of gas turbines is also discernible in total seasonal natural gas consumption: in 2013, gas used in July and August (to meet peak air conditioning demand) was 40% higher than gas used in January and February. Not surprisingly, new generating capacities in the United States have been dominated by gas turbines. In 2013, just over 50% (about 6.9 GW out of the total of 13.5 GW) of new capacity was in gas turbines, almost equally split between simple cycle and combined cycle arrangements; and in terms of total generating capacity in 2012, nearly a third was installed in natural gas-fueled gas turbines (20% in CCGTs), compared to 41% in steam turbines (including 29% in coal-fired plants and almost 8% in natural gas-fired stations), about 7.5% in hydroturbines and 5.5% in wind turbines (USEIA, 2013a).

Because of the rising demand and high price of high-quality materials used to make them, gas turbines have been getting more expensive to install, but CCGT still generates electricity far more cheaply than any other commonly deployed method. USEIA (2014g) projects the following levelized costs of new electricity generation for the entire systems entering operation by 2019 (with all values in $(2012)/MWh): about $95 for conventional coal, $96 for advanced nuclear (a highly arguable value), $103 for biomass, $80 for onshore wind (and $204 for offshore turbines), and $130 for solar PV—compared to $66 for conventional natural gas combined cycle and $64 for an advanced version of CCGT.

No other stationary prime movers combine so many advantages as do modern natural gas-fueled turbines: they have the most compact size and hence the smallest footprint of all electricity generators and hence can be easily installed within the boundaries of existing thermal power plants or industrial establishments; they can be inexpensively transported to those sites by trucks, barges, or ships; they are exceptionally reliable and relatively easy to maintain; their service is almost instantly available: they can reach full load within a few minutes and hence are perfect for covering peak load or other sudden fluctuations in demand; because they are air cooled, they (unlike steam turbogenerators) do not require any arrangements for water cooling; they are relatively quiet as silencers keep their noise below 60 dBA at the distance of 100; and in combined cycle arrangements, they have unrivaled efficiencies and hence also the lowest specific $CO_2$ emissions.

## 4.3   NATURAL GAS AS A RAW MATERIAL

All fossil fuels have been used not only as sources of heat and light but also as raw materials, most importantly as feedstock in chemical industries providing essential elements or compounds required for subsequent syntheses or processing. Because it is a mixture of the lightest alkanes, natural gas has an inherent advantage as a feedstock for chemical syntheses. During the combustion, all alkanes contribute to the release of heat, but natural gas is not used as feedstock either in its raw form or after processing (when its composition becomes more uniform), but its constituents are separated and used for specific applications.

Methane is the most important input for the synthesis of ammonia ($NH_3$), for the production of hydrogen (for hydrocracking and hydrodesulfurization) and methanol ($CH_3OH$) and its derivates. Ethane ($C_2H_6$), the lightest of natural gas liquids, is converted into ethylene ($C_2H_4$, ethene) whose polymerizations yield the largest and the most valuable chain of synthetic products. Similarly, propane ($C_3H_8$, also used as portable fuel, most commonly for home heating, small stoves, and barbeques) is turned into propylene ($C_3H_6$, propene), the second most important compound to be polymerized after ethylene. Butane ($C_4H_{10}$) yields butylene ($C_4H_8$, butene), the basis for synthetic rubber, but it is also blended with gasoline and mixed with propane and is sold as LPG.

But the first major use of natural gas as a raw material was in the production of carbon black. Carbon black is a powdery or granular form of virtually pure (at least 97%) elemental carbon (unlike black carbon, or soot, which contains <60% C). Industrial use of carbon black is dominated by rubber industry, in the United States about 70% of the total as a reinforcing filler in tires, 10% in other automotive goods (belts, hoses), and 10% into molded and extruded rubber products; the remainder is used as a pigment in printing inks (now ubiquitously in photocopiers and laser print cartridges) and coating resins and films (ICBA [International Carbon Black Association], 2014).

Before WWII, natural gas was the preferred raw material for carbon black in the US: plants in or near oil fields took advantage of inexpensive natural gas associated with oil production to meet the rising rubber demand created by new automobile age. In 1920, 8% of the US natural gas production was used by carbon black industry, and by 1940, that share rose to almost 18% (Schurr and Netschert, 1960). The industry relied on the thermal black process, with the gas injected into a reactor that contained a limited supply of air and the ensuing thermochemical decomposition produced carbon and hydrogen-rich combustion gas; the gas was burned off to generate heat required for the decomposition; and the carbon was separated, densified, and processed to specifications by screenings or pelletizing. Once the gas became more valuable, the industry switched to oil: since the mid-1970s, most of the material (now about 8 Mt/year) has been derived from heavy aromatic oils by the oil furnace process, and by the year 2000 only one out of 23 US carbon black plants was using the natural gas-based carbon black process (Crump, 2000).

Hydrogen produced by steam reforming of methane ($CH_4 + H_2O \rightarrow CO + 3H_2$) now accounts for roughly half of all $H_2$ derived from fossil fuels (the other half comes mostly from steam reforming of liquid hydrocarbons and coal, while the electrolysis of water, done with inexpensive hydroelectricity, supplies less than 5% of the global demand). Hydrogen's principal uses are in crude oil refining, for cracking, dearomatization, and desulfurization (Chang, Pashikanti, and Liu, 2012). Hydrogen now has a role in processing almost 4 Gt of crude oil every year, and with the average demand being roughly 0.5% (or 60 m$^3$/t of oil) of the total refinery crude throughput, total consumption is on the order of 20 Mt $H_2$ a year. The rest of some 50 Mt on the global market is roughly split between chemical syntheses (producing methanol, polymers, solvents, and pharmaceuticals) and a variety of industrial applications ranging from glass making to food processing (unsaturated fatty acids are hydrogenated to produce solid fats), with a minor share going for liquid rocket fuel.

## 4.3.1 Ammonia Synthesis

No use of methane as a raw material is more important than being a dominant source of hydrogen for the synthesis of ammonia ($NH_3$), the world's most important nitrogenous fertilizer without whose applications the global agriculture would not be able to feed at least 40% of the current humanity (Smil, 2001). Nitrogen, phosphorus, and potassium are the three macronutrients whose adequate supply is essential for high crop productivity. Potassium is supplied simply as potash (KCl, potassium chloride), a relatively abundant mineral that requires just mining and crushing before application to crop fields. Phosphate mining is highly concentrated in a few countries (the United States, Morocco, China), and the mineral is reacted with acids to prepare compounds (superphosphates) that yield the nutrient in forms that are more readily available to growing plants.

In all traditional agricultures, nitrogen came only from the recycling of organic matter (manures, crop residues) and cultivation of leguminous crop (symbiotic with nitrogen-fixing bacteria). The nineteenth century saw a relatively brief and intensive exploitation of guano (accumulated bird droppings on some subtropical islands) and mining and rising sales of the first inorganic form of fertilizer nitrogen in the form of Chilean nitrates. By the beginning of the twentieth century, there were two other minor inorganic contributors (ammonium sulfate from coking operations and synthetic cyanamide), but a truly revolutionary breakthrough was achieved only in 1900 when Fritz Haber succeeded to synthesize ammonia from its element under high pressure and in the presence of a metallic catalyst. Carl Bosch of the BASF (at the time the country's largest chemical company) led the effort to convert Haber's bench demonstration to a commercial process, and the world's first ammonia plant began to operate in October 1913 (Smil, 2001; Figure 4.4).

This innovation removed what has been always the most common limit on crop yield in all areas receiving adequate precipitation or irrigation, the availability of reactive nitrogen. Obviously, the synthesis of

Figure 4.4   Fritz Haber and Carl Bosch.

ammonia from its elements requires affordable supply of the two constituent gases, nitrogen and hydrogen. The former has been obtained since the 1890s by liquefaction of air and separation of its most abundant constituent. BASF's first plant, as well as all other early enterprises built in Europe during and after WWI, based their hydrogen production on coal. The process began with generating CO-rich gas and then, in the presence of metallic catalysts, transforming CO and steam into $CO_2$ and producing hydrogen $(CO + H_2O \rightarrow CO_2 + H_2)$.

German engineers knew that naphtha, a mixture of light liquid hydrocarbons, would make a much better feedstock than coal, but it was expensive and not readily available in the post-WWI Europe. BASF had eventually (between 1936 and 1939) developed a new process of partial oxidation $(C_nH_{(2n+2)} + n/2O_2 \rightarrow nCO + (n+1)H_2)$ that made it possible to use almost any hydrocarbon regardless of its molecular weight; as a result, heavy fuel oils became common feedstocks for $NH_3$ synthesis, particularly in India and China (Czuppon et al., 1992). But because partially oxidized heavier hydrocarbons have lower H:C ratio than $CH_4$, their use as feedstocks requires larger inputs as well as larger facilities for handling larger volumes of CO and $CO_2$.

Methane, the lightest homolog of the alkane series with the highest H:C ratio of all hydrocarbons, is the most economic choice for steam reforming, with every molecule of $CH_4$ producing three molecules of $H_2$ $(CH_4 + H_2O \rightarrow CO + 3H_2)$. Although no natural gas was available in post-WWI Germany, Georg Schiller, a BASF engineer, discovered during the 1920s how to reform $CH_4$ in an externally heated tube oven using a nickel catalyst. The process was licensed to the Standard Oil of New

Jersey which began to produce hydrogen in its Baton Rouge refinery in 1931, but the first ammonia plant using steam reforming of methane was built only in 1939 (Hercules Powder Company in California).

Postwar expansion of ammonia synthesis in the United States and the USSR was based entirely on the steam reforming of natural gas. European synthesis remained coal based until the late 1950s, but then the rising imports of crude oil and the discovery of natural gas in Groningen and in the North Sea and the imports from Siberia brought a fast shift to hydrocarbon feedstocks. Elsewhere, only China kept on building a relatively large number of smaller-capacity coal-based plants. This dependence was further strengthened since the 1960s with the replacement of old reciprocating compressors by much more efficient centrifugal machines.

Until that time, ammonia plants had a separate reciprocating compressor for each of its parallel synthesis loops known as trains. These compressors were first powered by coke-oven gas or by steam, after WWI mostly by electric motors, and although efficient they were also expensive (both to install and to operate) and could not handle very large volumes of air, limiting the daily capacity of a single train to no more than $300\,t$ $NH_3$. A new design was introduced for the first time in 1963 by M.W. Kellogg Company of Houston, now part of Kellogg Brown and Root, a leading engineering, construction, and service company that has been involved (in the licensing, design, or actual building) of roughly half of the global ammonia capacity (KBR, 2014).

In Kellogg's new plant, all compression needs were supplied by a single centrifugal machine powered by steam turbines, with the steam produced very efficiently by combustion of natural gas. This change—using natural gas as both the feedstock and the fuel energizing the process—allowed integration of the plant's energy needs, maximized the overall energy conversion efficiency, and made it possible to produce much larger volumes of $NH_3$ at a much lower cost. While the pre-1963 plants operated typically with pressures of $30$–$35\,MPa$ in the synthetic loop, the latest ammonia plants operate with pressures below $10\,MPa$. Capacity of the first M.W. Kellogg single-train plant in Texas City was $600\,t$ $NH_3$/day, soon afterward the first plant with daily output of $1,000\,t$ came, and more than two dozen similar facilities were at work by the late 1960s.

During the 1970s, the largest order for these new plants came from China, driven by the existential need to increase the country's stagnating crop production (Smil, 2004). By the end of the twentieth century, more than 80% of the global $NH_3$ synthesis capacity relied on steam reforming of methane, all large plants built since the year 2000 are natural gas based, and the largest single-train plants under construction now having daily output capacities of 4,000 t. But the sequence of ammonia synthesis has not changed: production of feedstock gases is followed by a shift reaction that greatly reduces the volume of CO and yields more $H_2$ and $CO_2$, then the two carbon oxides are removed, and pure $N_2$ and $H_2$ undergo high-pressure catalytic synthesis.

Natural gas delivered by a pipeline is first purified by removing traces of $H_2S$ and particulate matter, and then it is preheated and compressed to the reformer pressure and mixed with superheated steam. Its primary reforming (inside heated steel alloy tubes packed with nickel catalyst) yields a mixture of $H_2$ and CO $(CH_4 + H_2O \rightarrow CO + 3H_2)$, and it is the largest energy user in any ammonia plant. The gas is then led to the secondary reformer (a cylindrical vessel filled with a suitable catalyst) where the unconverted $CH_4$ (between 5 and 15% of the initial volume) is oxidized to yield $CO_2$ and water. The resulting gas contains 56% of hydrogen, 23% of nitrogen, 12% of CO, 8% of $CO_2$ and less than 0.5% methane as well as residual water. After cooling, a catalytic shift reaction produces more hydrogen $(CO + H_2O \rightarrow CO_2 + H_2)$, and all remaining $CO_2$ is removed by adsorption using either aqueous ethanolamine solutions or pressure-swing absorbers filled with microporous aluminosilicates (zeolites). Any residual traces of CO or $CO_2$ are eliminated by catalytic methanation $(CO + 3H_2 \rightarrow CH_4 + H_2O$ and $CO_2 + 4H_2 \rightarrow CH_4 + 2H_2O)$.

Synthesis gas (74% $H_2$, 24% $N_2$, 0.8% $CH_4$, and 0.3% Ar) is then compressed (6–18 MPa depending on the process), heated (to 400–450°C), and converted to ammonia $(N_2 + 3H_2 \rightarrow 2NH_3)$ in the presence of catalysts (formerly all based on magnetite $(Fe_3O_4)$ and promoted with $Al_2O_3$, KCl, and Ca, more recently also with ruthenium). The exit gas contains 12–18% $NH_3$; ammonia is refrigerated and stored; and the unreacted gas is compressed once again and led back into the converter. Besides America's KBR, other major licensors, consultants, and builders of ammonia plants are the Danish Haldor Topsøe, Germany's ThyssenKrupp Uhde GmbH, and the Swiss Ammonia Casale.

The worldwide switch to natural gas as fuel and feedstock for ammonia synthesis has transformed the industry. Its post-1950 growth has not been steady. Steady gains continued until the late 1980s, driven first by expanding crop production in the United States and Europe (both with growing populations and higher demand for animal foods), then by the nitrogen requirements of high-yielding rice and wheat cultivars of Asia's green revolution (Smil, 2000). The global output of ammonia (with about 80% of it destined for fertilizers, the rest for a variety of further chemical syntheses) rose from just below 5 Mt in 1950 to about 50 Mt in 1975. The USSR displaced the United States as the world's largest producer, and in turn, it was surpassed by China. Then the record output in the year 1989 was followed by a pullback during the early 1990s (largely due to declining output in post-Soviet states and reduced demand in Western countries brought by more efficient use of nitrogenous fertilizers) and then came renewed expansion.

The global capacity of ammonia plants surpassed 180 Mt in 2010 and it will be about 230 Mt in 2015 (FAO, 2011), and the actual output of ammonia reached 159 Mt in 2010 and 170 Mt in 2013 (equivalent of 140 Mt N), with nearly 56 Mt in China, almost 15 Mt in India, 12 Mt in Russia, and 10.6 Mt in the United States (USGS [United States Geological Survey], 2014). This means that ammonia production has been one of the world's two most important chemical syntheses when measured by the total output. In mass terms, the production of sulfuric acid and ammonia has been almost identical, but ammonia's lower molecular weight (17 vs. 98 for $H_2SO_4$) means that the gas has been the most important product when compared in terms of synthesized moles. Large-scale natural gas-based syntheses pioneered in the United States of the 1950s and 1960s have left their lasting imprint on the worldwide industry: the largest plants are located near major source of natural gas, and they are often integrated with production of more complex solid or liquid nitrogenous or mixed fertilizers.

Largest US plants (with annual capacities in Mt) are in Kenai in Alaska (0.63), Donaldsonville in Louisiana (four plants totaling 2.04 Mt), and Enid and Verdigris in Oklahoma, each nearly 1 Mt (IFDC [International Fertilizer Development Center], 2008). In Canada, they include Redwater and Carseland in Alberta, but the Western hemisphere's largest ammonia plant concentration is on the western coast of Trinidad in Point Lisas where the island's offshore gas fields feed 11 plants with the combined annual capacity of 6 Mt in 2013. Saudi plants are concentrated in al-Jubail on the Persian Gulf near the giant hydrocarbon fields of the country's Eastern Province, and the world's largest new plant (1.2 Mt/year capacity) will be completed in 2016 in Ras al-Khair, just north of al-Jubail. Russia's largest concentration of ammonia plants (seven facilities with the total annual capacity of 3.15 Mt) is in Tolyatti on the Volga, close both to the country's major (now declining) hydrocarbon basin (Volga–Ural) and a large hydrostation (Kuybyshevskaya on the Volga).

Ammonia has the highest nitrogen content (82%) of all fertilizers, but storage and applications of anhydrous ammonia require special equipment (tanks, hollow knives injecting ammonia into soil) and have been largely limited to the United States and Canada. Various ammonia solutions may be preferable, and because urea ($CH_4N_2O$), containing 45% N, is the most concentrated solid nitrogen fertilizer that is easy to store and apply, it has become the world's leading source of crop nitrogen, mainly because of its dominance in rice-growing Asia. Other common choices include ammonium nitrate ($NH_4NO_3$) and ammonium phosphate ($NH_4H_2PO_4$).

Gas-based synthesis is the least expensive option: plants based on the other two feedstocks (heavy oil and coal) cost 1.4–2.4 times as much as to build, 20–70% more to operate, and need 30–70% more energy. Switching to methane from the original coke-based synthesis was a major reason for impressive energy savings (Smil, 2001). Energy requirements of ammonia synthesis include fuels and electricity used in the process and energy embodied in the feedstocks. The first commercial operation in Oppau began with more than 100 GJ/t $NH_3$ in 1913, and during the late 1930s, coke-based plants consumed around 85 GJ/t $NH_3$. During the 1950s, natural gas-based plants using low-pressure reforming and reciprocating compressors required between 50 and 55 GJ/t $NH_3$, and by the early 1970s, when high-pressure reforming and centrifugal compressors became common, the rate was reduced to just 35 GJ/t $NH_3$.

By 1980, many plants required just 30 GJ/t $NH_3$, soon afterward new plant designs by M.W. Kellogg and Krupp Uhde reduced the rate to just below 29 GJ/t $NH_3$, and the best current performances are 27–28 GJ/t $NH_3$ based on natural gas, only about 30% higher than the stoichiometric minimum of 20.9 GJ/t (Worrell et al., 2008). Obviously, average global performances have been considerably less efficient (mainly due to higher energy intensity of plants that still rely on reforming heavier hydrocarbons or coal) as they declined from around 80 GJ/t $NH_3$ in 1950 to just over 50 GJ/t in 1980, to 45 GJ/t by the year 2000, and, according to the IEA, to a global weighted mean of 41.6 GJ/t in 2005 (ICF International, 2007).

## 4.3.2 Plastics from Natural Gas

The age of plastics depends heavily on two heavier hydrocarbons present in natural gas, above all on ethane, the second alkane in natural gas and the most voluminous component of natural gas liquids, and on propane. These constituents of natural gas provide the monomers for the production of the three dominant polymers—polyethylene (PE), polypropylene

(PP) and polyvinyl chloride (PVC)—whose consumption now accounts for roughly three-fifths of the global market for synthetic materials. Ethylene is also produced, more expensively, by cracking liquid naphtha obtained by the distillation of crude oil, and the research is advancing on methods to derive ethylene from biomass feedstocks—but the dominance of inexpensive ethane-based ethylene will not be dislodged easily.

Steam cracking of ethane produces ethylene, and after purification, it is catalytically converted to different kinds of PE. As already noted (in Chapter 2), ethane concentrations (by volume) range mostly between 2 and 7%, with extremes from 1 to 14% (Groningen 2.0%, Hugoton, TX 5.8%, Hassi R'Mel 7%, Agha Jari, Iran 14%), natural gas processing reduces high ethane level to concentrations prescribed by pipeline operators, and the separated ethane become a valuable feedstock. The only time this sequence presents problems is when new gas with high ethane levels is produced in regions that lack ethane-based industries. That has been the case in parts of the Marcellus shale in Pennsylvania where some wells produce gas with up to 16% of ethane (Martin, 2010).

About half of the world's ethylene production is polymerized: polymerization (conversion of small molecules, monomers, into long chains or networks of molecules, polymers) produces a variety of PE, the world's most important group of plastic. Polymerization process was first patented in 1936 by the British ICI, and commercial production of low-density polyethylene (LDPE) began in September 1939. High-density polyethylene (HDPE) was introduced during the 1960s; its density is only a bit higher than that of LDPE ($0.96 \, g/cm^3$ compared to $0.93 \, g/cm^3$), but its melting temperature is 20°C higher (135°C vs. 115°C), and its tensile strength is more than three times as much as for LDPE. During the 1970s came the commercial synthesis of linear low-density polyethylene (LLDPE) that is stronger than LDPE and can be turned into thinner films.

Global production of ethylene is now on the order of 160 Mt/year, with the Middle East (above all Saudi Arabia, taking advantage of its NGL supply), the United States, and China being the larger suppliers. Extrusion, molding, casting, and blowing turn different kinds of PE into a huge array of products for both ubiquitously visible and hidden uses. Thin PE films are used for shopping and garbage bags, food bags and wraps, and cover sheets (now common for cultivating vegetables); LDPE makes huge impact-resistant water tanks as well as soft bubble wraps; and LLDPE is used as frozen food bags, for heavy-duty liners, for pools, and for geomembranes. Leading hidden uses include house wraps, water pipes, insulation for electrical cables, and materials for knee and hip replacements. Given their widespread use, it is fortunate that PE products (triangle symbols 2 for HDPE, 4 for LDPE) are the most commonly recycled plastics.

Ethylene is also the starting material for the production of PVC, the second most common plastic worldwide. Catalytic reaction of ethylene and chlorine (released by electrolysis of NaCl) produces ethylene dichloride whose cracking results on vinyl chloride monomer that is turned into white PVC powder by polymerization. Although the global PVC output is lower than the PE production, the material is even more ubiquitous: wire insulation, sewage pipes, virtually every disposable item in a hospital (tubing, bags, containers, trays, basins, gloves), window frames, house siding, toys, and credit cards to name just a few common categories. But unlike PE, PVC has been often seen as a health hazard and a major environmental risk (Smil, 2013a).

The third in the trio of dominant plastics is PP produced since the mid-1950s by cracking propane and then by polymerizing propylene ($C_3H_6$, propene). PP is flammable and gets degraded by UV radiation, but it has many ubiquitous uses (above all various food containers, bottles, garbage cans, and pipes) that overlap with PE, but its combination of low density, fairly good strength, high melting point and resistance to acids and solvents makes it a preferred material for high-temperature uses (most commonly many items in hospitals that require sterilization) and heavy-duty applications (industrial pipes, container lids). PP is also used for carpets and outdoor fabrics and in several countries (including Canada) for new, more durable banknotes.

Oxidative catalytic dehydrogenation of butane produces butadiene ($C_4H_8 + \frac{1}{2}O_2 \rightarrow C_4H_6 + H_2O$) whose polymerization yields synthetic rubber: polybutadiene is highly resistant to wear, and that makes it an excellent candidate for the production of vehicle tires, and it is also used as a coating in electronics and in making a variety of elastic items. Polybutadiene rubber is highly elastic, it is often blended with natural rubber, and it now accounts for about a quarter of the global output of synthetic rubber. The most common (and the most affordable) kind of synthetic rubber is a copolymer of styrene and butadiene, while polyisoprene rubber shares the chemical structure with natural rubber but lacks its other ingredients and hence does not equal its quality.

Pentane is in a different category of raw materials than are the three lighter alkanes: rather than getting transformed by catalytic reactions into another compound, it is used a blowing agent in the production of expanded polystyrene (PS), a superior insulation material made of 90–95% of PS and 5–10% of pentane. PS beads are first expanded roughly 40 times by a small amount of steam-heated pentane; then the PS is cooled down and kept for up to 2 days in a holding tank before its reheating achieves the final expansion producing a material made up of as much as 98% of air. Pentane is also used as a blowing agent in making polyurethane foams.

### 4.3.3   Gas-to-Liquid Conversions

Industrial gas-to-liquid (GTL) conversions have followed two main pathways: methanol synthesis producing an alcohol and Fischer–Tropsch (F–T) synthesis producing synthetic crude oil (Olah, Goeppert, and Prakash, 2006; de Klerk, 2012; Brown, 2013). They both start with synthetic gas (syngas) made by catalytic steam reforming of a carbon-rich feedstock, solid, liquid, or gaseous. The starting raw material can be biomass, and until the 1920s, all commercial production of methanol was wood based; hence, its common name is wood alcohol. F–T production of gasoline by Nazi Germany was based on brown coal, and South Africa's production of gasoline and diesel (starting in 1955 in Sasolburg) was based on hard coal. Natural gas has become a dominant feedstock in the United States and in the Middle East because of its abundance and low cost.

Methanol ($CH_3OH$, methyl alcohol) is produced directly from synthetic gas (a mixture of $H_2$, CO, and $CO_2$) whose production from methane has been already described earlier in this chapter when dealing with ammonia. Subsequently, an exothermic catalytic reaction combines the gases to make methanol: $CO + 2H_2 \rightarrow CH_3OH$ and $CO_2 + 3H_2 \rightarrow CH_3OH + H_2$. About 30% of global methanol production (roughly 50 Mt in 2013) is used to make formaldehyde (widely used in wood, pharmaceuticals, and automotive industries), and another 30% goes for making acetic acid (mainly for adhesives and paints), methyl chloride (for silicones), and dimethyl terephthalate used for recyclable plastic bottles (Methanex, 2014). Production of methyl tertiary butyl ether (MTBE, added to gasoline to prevent knocking and raise octane rating) used to be a large consumer of methanol but the compound was outlawed in the United States. The most important energy use is for fuel blending: adding methanol to gasolines claimed about 12% of the total output in 2013 and the use is expanding strongly. I will return to this use in Chapter 7.

Production of methanol is the simplest GTL conversion: a much more complex procedure can turn the simplest alkane to high-purity liquid hydrocarbons, including gasoline, kerosene, and diesel fuel. These conversions are using F–T process first commercialized in Germany (Weil, 1949; Schwerin, 1991). Franz Fischer, Hans Tropsch, and Helmut Pichler developed the process between 1923 and 1926, and IG Farben built a prototype plant in 1926. F–T process starts with the splitting of CO and results in the formation of liquid hydrocarbons, mostly alkanes—$(2n+1)$ $H_2 + n$ CO $\rightarrow$ $C_nH_{(2n+2)} + n$ $H_2O$—but it also produces some alkenes and oxygenated hydrocarbons. Iron- and cobalt-based catalytic reactions produce liquids that have no sulfur and heavy metals (hence their combustion produces very little air pollution) and that can be hydrocracked into a variety of fuels (LPG, gasoline, jet fuel, naphtha, diesel) and lubricants, waxes, and white oils.

During WWII, Germany's coal-based GTL reached, in 1943, as much as 124,000 barrels per day (bpd), but then the allied bombing reduced the output to just 3,000 bpd by the fall of 1944 (Schwerin, 1991). The synthesis was abandoned after WWII as imports of inexpensive Middle Eastern and then Russian crude oil made such ventures uneconomical. Several small American plants built after WWII were closed by 1953, but in 1955, South Africa's Sasol began to produce hydrocarbons using large domestic coal deposits, first to lower dependence on imports and then to maintain domestic supply when the country's imports were embargoed due to its apartheid policies (Sasol, 2014). In 1997, PetroSA Mossel Bay plant, using natural gas from fields some 120 km offshore, became the world's first GTL using a process developed by Sasol for its planned ventures in Qatar and Nigeria. Sasol's first overseas plant, Oryx GTL in Qatar (daily capacity of 33,700 barrels of liquids), began to operate in 2007, and the company has several plans for facilities in North America.

Among large oil and gas companies, Shell has been the one most committed to the development of large-scale GTL (Shell, 2014). Since the 1970s, it has amassed more than 3,500 patents for all stages of the process, and it used its experience from its first GTL plant in Bintulu in Malaysia (which began to operate in 1993 with a current daily capacity of 14,700 barrels) to build the world's largest project in Ras Laffan Industrial City in Qatar: Pearl GTL entered full production by the end of 2012, and Shell is the operator of the plant developed under a production and sharing agreement with Qatar Petroleum. Pearl GTL uses natural gas from the world's largest gas field (North Dome): 22 wells produce maximum of 1.6 billion cuft/day to be converted, by proprietary Shell Middle Distillate Synthesis, into 140,000 barrels of virtually sulfur-free, highly biodegradable, and nearly odorless products (gas oil, kerosene, naphtha, and paraffins) and 120,000 barrels of natural gas liquids and ethane (Shell, 2014; Figure 4.5).

**Figure 4.5** Qatar Shell Pearl GTL. Reproduced with permission from Photographic Services, Shell International Limited.

Pearl GTL was to be profitable at oil prices between $50 and 70/ barrel, but after it began to operate, prices were fluctuating at around $100/barrel making its operation quite rewarding—but this has not resulted in a worldwide stampede to build more large GTL projects: high capital cost of these facilities is a key deterrent. In the late 1990s, the capital cost for GTL plants was estimated to be as low as $20,000/ bpd, Shell's Pearl GTL costs $140,000/bpd, and Nigerian Escravos (Sasol–Chevron) is estimated at $180,000/bpd (de Klerk, 2012). Escravos was designed to eliminate large-scale flaring of natural gas in the Niger Delta by setting up a 33,000 bpd GTL facility (processing 325 Mcf/day) based on the Oryx experience as a collaboration of Sasol, Chevron, and the Nigerian National Petroleum Corporation.

But the project's completion has been postponed several times before the start-up in mid 2014 (Chevron, 2014), and its cost (close to $10 billion) is roughly eight times of a similarly sized Oryx which required $1.2 billion (Sasol, 2014). And (a development not anticipated even just a decade ago) petrochemical producers in the Gulf countries are now experiencing regional shortages of natural gas and are forced to use liquid feedstocks (Horncastle et al., 2012). In any case, perhaps the best way to illustrate how far the GTL projects have to go in order to make a real difference to the global liquid fuel supply is to note that the aggregate capacity of all GLT projects operating in 2014 (<400,000 bpd) was equal to less than 0.5% of the global liquid hydrocarbon extraction.

The ultimate goal of natural gas conversion is to commercialize new methods that would transform methane into cheap feedstocks. Perhaps the most publicized laboratory demonstration achieved selective conversion of methane to methanol at temperatures around 200°C using platinum bipyrimidine complexes as catalysts (Periana et al., 1998). Solid catalysts would be preferable (Palkovits et al., 2009), but as is often the case with laboratory demonstrations, scaling up this pathway as the basis of commercial production is another matter (Service, 2014). Other intriguing proposals envision enzymatic bioconversion of methane to liquid fuels (Conrado and Gonzalez, 2014) and direct, nonoxidative conversion of methane to ethylene, aromatics, and hydrogen (Guo et al., 2014). Again, challenges inherent in scaling such processes to produce millions of tons of feedstocks a year remain immense.

# 5

# Exports and Emergence of Global Trade

Coal exports are well documented already in medieval Europe, starting with the English sales during the fourteenth century: before its end, dozens of boats were engaged in regular shipment of the fuel from Newcastle to European ports on the coasts between northern France and Denmark (Daemen, 2004). Crude oil was traded internationally almost from the very beginning of its commercial extraction, and so were the numerous refined oil products, including such important nonfuel items as lubricating oils, waxes, and asphalt. By the end of the nineteenth century, the two largest oil producers, Russia and the United States, were also the largest exporters of petroleum products, mainly to Western Europe.

By 1950, even before large-scale oil imports to Europe and Japan, 27% of the world's crude oil production was internationally traded, and that share rose to 36% by 1960 and to nearly 55% by 1974, just after the OPEC's action had quintupled the world oil price (UNO [United Nations Organization], 1976). In contrast, large-scale international trade in natural gas was nonexistent or very limited as long as only a few countries were substantial producers (recall that even as recently as 1970, there were only five countries, the United States, Canada, Netherlands, Romania, and the USSR, producing annually more than $20\,Gm^3$ of natural gas), and once the supply had expanded, development of substantial gas trade required first major capital investments in transcontinental pipelines (for trade within North America or Europe) and

*Natural Gas: Fuel for the 21st Century*, First Edition. Vaclav Smil.
© 2015 John Wiley & Sons, Ltd. Published 2015 by John Wiley & Sons, Ltd.

then in expensive infrastructures to liquefy, transport, and regasify gas (for imports from other continents).

But once these two processes got underway, the progress was relatively rapid. In 1950, the international gas trade amounted to a small fraction of 1% of the extracted fuel, and even by 1960, the only major export stream was the western Canadian gas (about 2.6 Gm$^3$) going to the US Midwest. But by the century's end, nearly 21% of all extracted natural gas was sold abroad, and by 2012, 31% of all marketed natural gas was traded internationally (21% by pipelines, 10% by liquefied natural gas (LNG) tankers), compared to 64% of crude oil and about 16% of bituminous coal (WCA (World Coal Association, 2014). As a share of the total output, natural gas is thus now traded about twice as much as coal and about half as much as crude oil, but a better comparison is in terms of energy equivalents: in 2012, natural gas export reached about 36 EJ, while crude oil exports were 3.2 times larger at 115 EJ and coal exports were about 31 EJ, or 15% lower than the international gas sales.

Compared to crude oil (and even more so to the sales of refined oil products), the natural gas trade has been also much less diversified. Virtually every one of the world's roughly 200 countries (including major crude oil producers) is importing some refined oil products, and 16 countries are major crude exporters (shipping more than a million barrels a day): Saudi Arabia, Russia, the UAE, Kuwait, Nigeria, Iraq, Iran, Qatar, Angola, Venezuela, Norway, Canada, Algeria, Kazakhstan, Libya, and Algeria. In addition, 20 other countries sell more than 100,000 barrels of crude a day (CIA, 2014). In contrast, the total number of natural gas importers remained limited until the 1960s when the Dutch and Algerian gas began to reach the European market and widened further after 1973 when the first Soviet gas exports reached both East and West Germany, while the number of exporters began to increase only during the 1990s with the expansion of LNG trade.

In 2012, dozen countries were major natural gas exporters by pipeline led by Russia and followed (in volume order) by Norway, Canada, and the Netherlands. These four countries accounted for 61% of all exports by pipeline. The next four largest exporters were Turkmenistan, Algeria, Qatar, and Bolivia. When the exports of natural gas (petroleum gases in the terminology of the Standard International Trade Classification, item 3414) are expressed as the shares of total foreign earnings, the countries that were most dependent on such sales in the year 2010 were Bolivia (41%), Turkmenistan (40%), Algeria (20%), Norway (16%), and Russia with nearly 9% (Hausmann et al., 2013). As I will detail later in this chapter, the trade in LNG was slow to expand, but its total volume has reached about 50% of the amount exported through pipelines ($330\,Gm^3$ compared to $705\,Gm^3$ in 2012) with 18 LNG-exporting countries led by Qatar (accounting for nearly a third of the total 2012 volume), with Malaysia the distant second (<10% of all LNG trade) followed by Australia, Nigeria, and Indonesia (each on the order of 8% of the global total).

In this chapter, I will look in some detail at the two major continental natural gas markets: at North America with its now rapidly changing supply and demand with imports accounting for about 18% of the global total and at the world's largest, and increasingly interconnected and integrated, market that links the giant fields of Asia (Siberian Russia, Central Asia, and the Middle East) with the importers in Europe and the Far East and that involves about 75% of global natural gas pipeline trade. The chapter's last segment will be devoted to the slow evolution of LNG industry and to the assessment of its current state and near-term prospects: will it become a truly global endeavor akin to the trade in crude oil and refined oil products (with scores of importers) or will its importance remain limited to only a few regions?

## 5.1 NORTH AMERICAN NATURAL GAS SYSTEM

Origins of North American natural gas trade go back to the 1890s. Canada began exporting natural gas from Welland, Ontario, in 1891 to Buffalo in Upstate New York (at that time a major US industrial center), and in 1897 a pipeline under the Detroit River brought gas from the Essex field to Detroit, Michigan, and also to Toledo, Ohio. Declining output of the Essex field soon led the Ontario government to prohibit exports of both gas and electricity. Canada had actually become a small-scale importer of the US gas, and the situation changed only when major post-WWII hydrocarbon discoveries in the Western Canada Sedimentary Basin (in Alberta and British Columbia) transformed Canada into a major crude oil and natural gas exporter, with all of the gas going to the United States.

In 2012, Canada had just over 1% of the world's proved gas reserves, but until 2011, it was the world's third largest producer of natural gas (behind the United States and Russia), and only in 2012, it was surpassed (by <3%) by Iran, the country with the world's largest (just over 18% of the total) conventional gas reserves. Three factors explain the great disparity between Canada's modest natural gas endowment and its large annual gas extraction: domestic needs for heating, with Canada averaging recently 3,500–4,500 heating degree days, compared to 2,000–2,500 in the United States (CGA [Canadian Gas Association], 2014); energy-intensive categories of industrial production (petrochemicals, synthetic fertilizers, ferrous and color metallurgy); and the country's proximity to the huge US market that has for decades guaranteed a rising demand for gas imports.

As a result, Canada has been sending large shares of its gas extraction—30% in 1980, 38% in 1990, 49% in 2000, 41% in 2010, and 51% in 2013—to the United States, making it the world's second largest gas exporter (after Russia) until 1980 when it was surpassed by Norway and the third largest exporter until 2009 when it was surpassed by Qatari shipments of LNG. Canada has been the source of all but a small fraction of natural gas imported to the United States. Imports from Mexico have been small and sporadic (up to $2.9\,Gm^3$ in the early 1980s, none later in the decade, and no more than $1.5\,Gm^3$/year since the year 2000), and imports of LNG (started in 1985) remained below $5\,Gm^3$ until the year 2000 and peaked at almost $22\,Gm^3$ in 2007.

This means that Canada supplied 99.5% of all imports in 1975, 94% in 2000, 82% in 2007, and, once again, 97% in 2013. Shares of Canada's natural gas in the overall US consumption were just over 5% in 1975, 15% in 2000, and 16% in 2007 with, naturally, much higher contributions in main import markets in the Pacific Northwest (Seattle), the Midwest (Minneapolis, Milwaukee and Chicago), and the Northeast (Boston). And a comparison in absolute terms shows even better the magnitude of that trade: when Canada's natural gas exports peaked at $107\,Gm^3$ in 2007, they were larger than the total gas production of all but six leading countries, and their volume was larger than the current natural gas production of Saudi Arabia.

Obviously, such a level of trade has required enormous pipeline infrastructure (USEIA [US Energy Information Administration], 2008; NEB, 2014a) and some thirty border crossings (Figure 5.1). Among the many tangible signs of the economic integration of Canada and the United States, this one is perhaps the least known and least appreciated by the public: unlike crude oil whose extraction from oil sands and increasingly common transportation by railcars (and more common accidents) attract plenty of public attention, the multibillion natural gas trade takes places quietly out of sight, the only common evidence of it being small signs indicating the routes of buried pipelines.

The first Canadian natural gas pipeline to the United States was built by the Westcoast Transmission Company (now operating as Spectra Energy); it began the construction of a 60 cm diameter pipeline from

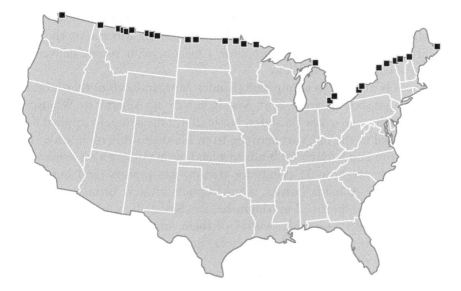

Figure 5.1  Canada–US natural gas pipeline crossings.

Taylor in northeastern BC to the BC–US border in 1955, and deliveries began in 1957 (CEPA [Canadian Energy Pipeline Association], 2014). In 1956 (after 6 years of planning and often contentious political, financing, and regulatory process), TransCanada Pipelines began its construction of a transcontinental line from Alberta–Saskatchewan border to Toronto (TransCanada, 2014). Alberta gas reached Winnipeg in 1957, and after the most difficult section through the rocks of the Canadian Shield, the construction was finished in October 1958. The line was extended to Montreal in 1976, and its length (3,500 km) was surpassed only by the Soviet line from Western Siberia. TransCanada Pipeline owns the most extensive network of natural gas pipelines in North America (13 major systems and nearly 60,000 km of trunk lines of which about 41,000 km are in Canada).

In 1981, Foothills Pipe Lines brought gas from Central Alberta to the US border, and in the Westcoast Energy operates a system that extends from parts of Yukon and Northwest Territories through Alberta and British Columbia where it joins the US trunk lines near Huntingdon and supplies gas to the US Northwest market (in recent years mostly between 20 and 30 Mm³/day). Spectra Energy owns nearly 5,700 km of pipelines that move the western gas east to Canadian and US markets, and it also owns the Maritimes and Northeast Pipeline that brings natural gas from Nova Scotia to the Eastern United States. Alliance Pipeline (about

3,700 km, completed in 2000) brings liquid-rich natural gas from northeastern British Columbia and northwestern Alberta to the US Midwest, specifically to a hub near Chicago, with recent deliveries clustered around 50 Mm$^3$/day.

There is also a special pipeline (Kinder Morgan Cochin) that moved natural gas liquids from Fort Saskatchewan in Alberta to Windsor, Ontario, through seven US states, but in 2012, the company applied for the reversal of the line's western leg from Illinois to Alberta in order to bring more condensate for bitumen blending in Alberta's oil sands. In total, there are now 31 points for natural gas exports (and imports) along the US–Canada border: a single point in New Brunswick, 3 in Quebec, 10 in Ontario, 2 in Manitoba, 7 in Saskatchewan, 6 in Alberta, and 2 in British Columbia, with 5 of them (at crossings to Montana, Idaho, North Dakota, and Minnesota) accounting for 70% of all US pipeline imports.

Even as it had been importing more natural gas from Canada, the United States remained an exporter of gas to Mexico. After the first global oil price spike in 1973–1974, these exports were cut by nearly 90% (from about 400 Mm$^3$ in 1973 to just 46 Mm$^3$ by 1983), but afterward, their almost uninterrupted rise brought them to nearly 3 Gm$^3$ in the year 2000 and to a record high of 18.6 Gm$^3$ in 2013. And, concurrently, exports to Canada became even larger. There were always small seasonal gas shipments from the United States to Canada, running on the order of 2–3 Gm$^3$, but this has changed with the sudden availability of American shale gas. In 2011, as Canadian gas exports fell by 5%, the US sales to Canada rose by 27% to more than 26 Gm$^3$ and increased further to 27.5 Gm$^3$ in 2012 when they equaled a third of the US gas imports. As a result, annual throughput of several Canadian pipelines (particularly of the TransCanada Mainline) has decreased (to as low as 43% of capacity at the Iroquois transit point in Ontario during the first 6 months of 2012), and that resulted in increased operating tolls for the remaining customers (NEB, 2014a).

But such expense is negligible compared to Canada's decreased export earnings. During the 1990s, export prices were fairly steady (fluctuating mostly between $1.5 and 2.0/1000 cuft), but a subsequent rise brought them to the $6.83 in 2006 and 2007 and to the record level of $8.58 in 2008. Combined with the record level of shipments, this translated to earnings of about $25 billion in 2006 and 2007 and $30 billion in 2008. But the price was halved in 2009 ($4.14), and it was down to $2.79 in 2012 (adjusted for inflation the last time the natural gas price was that low was in the late 1990s). How much further will the Canadian export decline will depend on the progress of US shale gas extraction and its share exported as LNG.

Canada has more natural gas reserves ready to be developed: after a lengthy regulatory review, Canada's National Energy Board approved the construction of the Mackenzie Valley Pipeline to bring the Beaufort Sea Arctic gas to Alberta and then to the United States, but its prospects remain poor because of the surge of the US shale gas extraction and Alaska gas pipeline that has already greatly reduced the decades-long dependence of the United States on rising Canadian gas imports. Total US natural gas imports were steady until 1986 (fluctuating narrowly around $28\,Gm^3$), and then a nearly uninterrupted rise brought them to $130\,Gm^3$ in 2007 (USEIA, 2014h). This peak was followed immediately by a fairly rapid decline: by 2013, the imports were 37% lower, a shift with an obvious impact on Canadian sales. With new US pipelines bringing gas from Marcellus shale to markets in the US Northeast and the Midwest, it is very unlikely that Canadian gas exports will return to their peak levels anytime soon, but it is also highly unlikely that they will be soon reduced to marginal levels.

In any case, the change has been pronounced. The United States became the net importer of natural gas in 1958 when it purchased almost 1% of its total consumption. Post-1985 combination of stagnating domestic production and increasing dependence on imports led to projections of massive LNG imports during the second decade of the twenty-first century. But immediately after the import share reached its historic peak in 2007 at 16.3%, the trends began to reverse with speculations about the future level of the US exports and the magnitude of ensuing earnings entirely displacing worries about the rising import dependence. Nothing illustrates this reversal better than the conversion of some of the US regasification facilities into liquefaction plants for exporting shale gas to Asia and Europe. I will take a closer, critical look at these speculations in the last segment of this chapter dealing with LNG and, once again, in the book's last chapter when I will try to answer the question about how far the gas will go.

## 5.2 EURASIAN NETWORKS

America's natural gas imports have been, primarily, the result of very high rates of per capita energy consumption that could not be entirely supplied by massive domestic production and had to be supplemented for decades by increasing shipments from Canada and, to a much lesser extent, by purchases of foreign LNG. In contrast, European natural gas imports have been a necessity, not an existential one but an unavoidable as long as the continent's countries wanted to improve their standard of living while reducing the environmental impact of high energy consumption: rise of natural gas imports had its obverse in falling coal combustion, now the dominant source of primary energy in a single EU country (Poland).

Europe's natural gas pipeline network has evolved largely in response to four major developments: discovery of Groningen gas, discovery of the North Sea gas, imports of Russian gas from Western Siberia, and increased imports of LNG from North Africa and the Middle East (Figure 5.2). Following chronologically the construction of all of the continent's major pipelines and charting the emergence of Europe-wide natural gas networks (now connected to Western Siberia, Central Asia, and Algeria) would be tedious: instead, I will describe the major lines of the system, including its latest extensions, as well as some of the most notable plans for its further expansion.

Europe, for decades considered devoid of any major hydrocarbon resources, was lucky as the discoveries of giant gas fields, first in the Netherlands and then in the North Sea, brought huge gas reserves within easy pipeline reach of the continent's three largest economies. But it was not enough to satisfy the rising demand, and the continent has developed a rising dependence on the export of Soviet and, after 1991, Russian natural gas (Clingendael, 2009; Högselius, Kaijser, and Åberg, 2010; Smeenk, 2010). European natural gas grid began to develop during the late 1960, with the sales of Groningen gas to Belgium and Germany in 1966 and to France in 1967 and with the sales of Soviet gas from Ukrainian fields (discovered before WWII, Shebelinka being the largest one) to Czechoslovakia, also in 1967 (called, a year before the Soviet invasion of the country, *Bratstvo*, Brotherhood).

In 1968, Austria was the first country outside the Soviet block to get the Soviet gas via a short spur from Czechoslovakia; West Germany followed in 1973 (in that year, East Germany was also connected), Italy in 1974, and France in 1976. All of these deliveries used the same trunk line completed in 1967, and although its capacity rose from less than $7\,Gm^3$ in 1973 to nearly $55\,Gm^3$ by 1980 (Gazprom, 2014), it was obvious that a new line will have to be built to connect giant fields of West Siberia with Central and Western Europe. Construction of 4,451 km long line from Urengoy to Uzhgorod station on the Ukrainian–Slovak border began in 1982. Construction of this Cold War pipeline was opposed by the United States because of its concern about the increased Soviet political leverage (CIA,

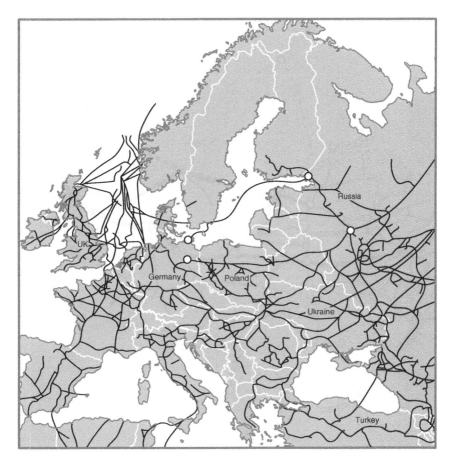

**Figure 5.2** European gas networks.

1981)—but Germany provided financing, and together with other European countries (led by France and Italy) and Japan, it also supplied the required pipe, pipe-laying equipment, and gas turbines for compressor stations. The line was built in stages and completed according to schedule in 1984 (Figure 5.3).

Parallel lines were added during the late 1980s, and in 1994, construction began on a more northerly line from Yamal to Germany. Unlike all the previous lines crossing the Ukraine, this line goes via Smolensk and Minsk and transports the gas to Germany across Belarus and Poland. The 4,190 km Yamal line began to operate in 1997, but its full design capacity of 32.9 Gm$^3$ was reached only in 2006. As already noted in the first chapter, yet another route to Germany was taken by the Nord Stream projects whose two parallel lines were laid between 2010

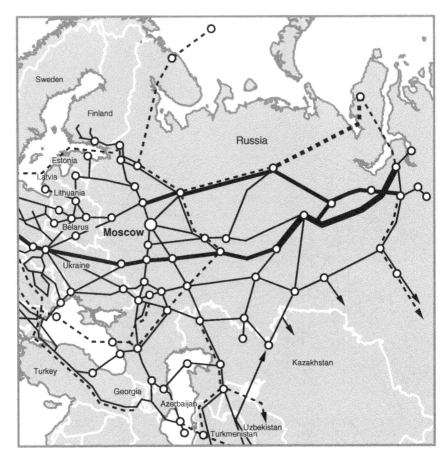

**Figure 5.3**   Russian export pipelines.

and 2012 on the floor of the Baltic Sea on their 1,224 km transit from Vyborg to Lubmin (Nord Stream, 2014; Figure 5.2).

The northernmost EU country to get the Russian gas is Finland, the southernmost one is Greece, but Russian gas also reaches Turkey: it got to Istanbul first via Ukraine, Moldova, Romania, and Bulgaria, but by 2003, a direct 1,213 km Blue Stream line was laid across the Black Sea from to Samsun on the northern Turkish coast and then to Ankara. By 2014, pipelines carrying the Russian gas westward had a capacity of 241 Gm$^3$ (142 Gm$^3$ via Ukraine, 38 Gm$^3$ via Belarus, 6 Gm$^3$ to Finland, and 55 Gm$^3$ for the Nord Stream), and the Blue Stream was designed for 16 Gm$^3$ (EEGA [East European Gas Analysis], 2014). Russian gas was sold to 23 European countries (spatially ranging from Greece to Finland and from Latvia to the United Kingdom) as well as to Turkey.

Actual gas deliveries to the EU countries, other European non-EU states, and Turkey rose from 130 Gm$^3$ in 2000 to 154 Gm$^3$ in 2005, stagnated before reaching a new record of about 161.5 Gm$^3$ in 2013 or 30% of those countries' consumption, with 16% of Europe's total supply now passing through Ukraine. South Stream pipeline would bypass Ukraine and deliver gas directly across the Black Sea to the Balkans, supplying first Bulgaria, Serbia, and Hungary and then linking to the existing network in Italy, and it would add another 63 Gm$^3$ of export capacity starting in 2016 (South Stream, 2014). In June 2014, the EU asked Bulgaria to stop work on South Stream as its construction conflicts with the Union's rules on public procurement and liberalization of energy markets, and in December 2014, Russia officially scrapped the project and decided instead to build a new pipeline to Turkey.

Gazprom, the descendant of the Soviet natural gas ministry that now accounts for about 75% of Russia's natural gas extraction, controls all Russian gas exports, but its long-held advantages—high profits and apparently unassailable dominance as the prime supplier of the EU needs—have been eroding due to high investment needed to keep up the output of its aging Siberian fields and bringing new fields on line (an effort requiring also new long pipeline or new LNG terminals), as well as due to the EU's effort to diversify its sources and reduce its dependence on the Russian gas. But realities make rapid diversification of supplies impossible, while politics (domestic and foreign) and business ties preclude many clear national strategies.

Dependence on Russian natural gas ranges from absolute in the east to very important as far west as Germany and Italy. In 2013, five EU countries (Finland, Latvia, Estonia, Lithuania, and Bulgaria) got all of their natural gas consumption from Russia; the shares were more than 98% in Slovakia, 86% in the Czech Republic, 84% in Poland, close to 70% in Greece and Hungary, and 42% in Germany (BP [British Petroleum], 2014a). Moreover, Gazprom points out that the consumption of the Russian gas as the share of the EU total supply has actually risen (to 28.3% in 2013, after falling from 25.3% in 2000 to 22% in 2010), that the company is a more reliable supplier than the nearest alternatives (Algeria or Libya), and that it can cover best any rising seasonal swings in daily gas deliveries.

Most importantly, there is no other supplier positioned so well to cover the continent's widening gap between extraction and demand that stood at about 300 Gm$^3$ in 2013 but is expected to rise to 400 Gm$^3$ by 2025 and to 45 Gm$^3$ by 2035; it now imports nearly half of its gas, but by 2030, it may have to buy 75% of it. This high degree of interdependence—the EU's high, and rising, dependence on imports and Russia's natural advantage to supply them—leaves no room for sudden dramatic shifts. As Gazprom's head of contract structuring and investment flows put it, EU–Russian gas trade is simply too deep and too comprehensive to fail (Komlev, 2014).

Moreover, Germany, the EU's leading economy and the largest importer of Russian gas, has its strongly pro-Moscow lobby led by the former chancellor Gerhard Schröder, now the chairman of the board of the Nord Stream. And given the strong German exports to Russia, many large companies (Siemens, Daimler, BMW) as well as scores of *Mittelstand* exporters are eager not to upset the existing trade arrangements. Gazprom, whose stock value was cut by 75% between May and October 2008 and recovered only about 20% of that loss by summer 2014 (Gazprom, 2014), has responded to European diversification moves (small volumes of LNG are now coming from Trinidad and Tobago, Qatar, and Nigeria) by strategic pricing.

When taking the price it charges to Germany, its largest buyer ($380/1000 m$^3$) as the standard, the United Kingdom gets about 15% discount but Switzerland has to pay 15% more, and Poland pays an almost extortionary price that is 40% higher (Eyl-Mazzega, 2013). On the other hand, as oil and gas exports account for 80–90% of Russia's foreign trade with many EU countries and gas alone makes about 10% of all Russian exports, Gazprom will have to be more flexible in order to keep this revenue stream from gradually declining, and it will have to do more to lessen concerns about the reliability of its deliveries.

Suggestions for improving the EU's bargaining power by applying competition law rigorously, by licensing (with conditions) all Russian gas imports, or by confronting Gazprom monopoly with a monopoly of a single central EU gas buyer are listed by Helm (2014) as a part of a credible European energy security plan, but their early adoption is unlikely. Even less likely is an early diversification of the EU's energy supply away from natural gas: most of the member countries see nuclear power as unacceptably risky, increased coal combustion would destroy the commitment to lower carbon emissions, and wind and PV electricity supply only marginal shares of overall primary energy demand (Smil, 2014).

Pipelines crossing Ukraine were not the only lines built during the Soviet era that became international lines after the dissolution of the USSR. During the 1970s and 1980s, lines were built to bring natural gas from giant fields of Turkmenistan (Shatlyk, Bayram Ali, Kirpichli) and Uzbekistan (Urta-Bulak, Dengizkul, Kandym) to the European part of the USSR and then to Europe via the *Soyuz* line, with a branch going northeast just west of the Aral Sea to supply the industrial areas in the southern Ural region. In the south, the Turkmen gas is transported eastward to Kyrgyzstan, Tajikistan, and Southern Kazakhstan. The other line that is now international transports Azeri Caspian Sea gas from Baku to Russia as well as to Armenia and Georgia.

Two recent developments have brought a fundamental shift to Eurasian gas exports (Figure 5.4). In 2009, Turkmen natural gas began to flow to China via 1,833 km long pipeline (diameter of 1.067 m) starting at the Uzbekistan border and ending in Horgos in Xinjiang. The second parallel line was completed in 2010, and they also receive gas from fields in Uzbekistan (long-term contract for 10 Gm$^3$) and Kazakhstan. Gradual increase of shipments should be the total volume of exported gas to 65 Gm$^3$ by 2020 (Gurt, 2014). And in May 2014, Russia and China finally concluded a long-contemplated deal (originally involving the West Siberian gas) to sell Russian gas to China. East Siberian gas from Chayanda field (in Sakha, formerly Yakutia, with reserves of 2.4 Tm$^3$,

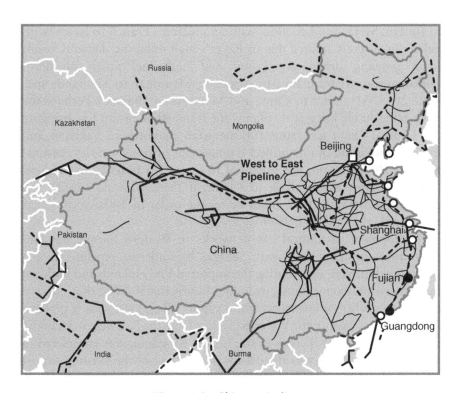

**Figure 5.4**   Chinese pipelines.

not yet in production) and Kovykta (west of the Lake Baikal, reserves of about 2 Tm³) will be the source, and the deliveries should start in 2018 (Itar-Tass, 2014a).

Annual volume of 38 Gm³ is not particularly large considering either the total Russian output or the huge Chinese demand, but the deal is to last 30 years and hence the total volume will surpass 1.1 Tm³ which is more than, for example, the total reserves of Azerbaijan or the Netherlands. Nevertheless, the CEO of Gazprom claimed that the deal will influence the whole gas market (Itar-Tass, 2014a), but the undisclosed price (between $350 and 380/1,000 m³) is no higher than the discounted long-term contracts signed with major European importers during 2013, and the construction of nearly 4,000 km pipeline to Vladivostok on the Pacific Ocean (to feed Russia's future LNG exports) will cost close to $50 billion, and the branch to China will add another $20 billion, totals that make the deal less appealing than the sales to Europe.

The rest of Asia is still far behind the emerging supernetwork of long-distance trunk lines connecting China with the Central Asia and Siberia. There are a number of shorter lines for selling gas to neighbors: Egypt to Jordan, Syria, and Lebanon, with an undersea branch to Israel (with deliveries severely limited due to Egypt's high domestic demand, result of a growing urban population and high fuel price subsidies); Turkmenistan to Iran; Iran to Armenia; Myanmar to Thailand; and, since 2013, Myanmar to China and Malaysia to Thailand. Perhaps the most remarkable large-scale plan is the trans-Afghanistan line bringing the Central Asian gas from Turkmenistan to northern Pakistan and India. Completion of this project remains (for obvious security reasons) particularly uncertain, but potential to export to India is, in the long run, no smaller than selling to China: if there were no supply and payment constraints, they could be on the order of 100 Gm³ by 2025 and double that a decade later.

Development of natural gas networks in Western Europe has had three aims: bringing the North Sea gas by undersea pipelines to the continental market; distributing the imported LNG inland; and building many new interconnections, including bidirectional links, that allow a greater flexibility of intracontinental supply. The oldest subsea pipeline for natural gas export, the Vesterled system from the Norwegian Heimdal Riser platform to St. Fergus in the United Kingdom, was completed in 1978, and the 1990s was the decade when most of the existing export lines went into operation: Europipe I and II in 1995 and 1999 (from Norwegian waters to Germany), Franpipe to Dunkerque in 1998, and

Zeepipe from the Troll processing plant near Bergen to Belgium (Subsea Oil and Gas Directory, 2014). Langeled, the region's longest (1,166 km) export pipeline taking gas from Nyhamma via Sleipner field to Easington in the United Kingdom, was finished in 2006.

Europe got its first connection with North Africa in 1983 when the Trans-Mediterranean pipeline (2,476 km) began to transport Algerian gas from Hassi R'Mel via Tunisia to Sicily and then across the center of the island and the Strait of Messina to Italian mainland and all the way to Slovenia. Thirteen years later came the exports of Algerian gas westward via Morocco and the Strait of Gibraltar to Spain, to Cordoba distance of 1,620 km. The gas is then moved to Madrid and northward to a low-capacity interconnection with France. By 2013, Europe also had 19 regasification terminals for natural gas imported (leaving aside small intra-European LNG exports from Norway) from North Africa, Nigeria, the Middle East, and Trinidad and Tobago. I will take a closer look at the development of Europe's LNG imports in the next section of this chapter when I will follow the industry's long evolution.

Europe has been looking beyond its four current major supply streams (Russia, North Sea, Groningen, and LNG) to pipeline imports from Azerbaijan, Turkmenistan, and even Iraq. The Nabucco pipeline was originally conceived to supply Azeri gas to Europe via Turkey, with Iraqi gas added later. The original plan was for 3,893 km line from Ahiboz in Turkey (where it would receive gas from South Caucasus line and an Iraq line) to Baumgarten in Austria, but after more than a decade of plans and negotiations and contracts, the line's fate remains uncertain. A competing project is the trans-Caspian line (capacity of 30 Gm$^3$/year) that would run from Turkmenistan to Azerbaijan and then across Turkey to Europe, completely bypassing Russia. Opportunities for further pipeline extension and grand integration of Russian, Chinese, Middle Eastern, and European networks would change radically should Iran join the ranks of normal countries, but there is no chance that now so deeply embedded (soon to be 40 years) Shiite theocracy would yield its control voluntarily and that Iran will emerge as natural gas export power rivaling Russia.

The unsettled state of the Middle East has been the major reason why there have been no serious plans to connect gas from the world's largest natural gas field to European markets by pipeline: a line from Qatar to the EU would have to cross Saudi Arabia and then either Kuwait and Iraq or Syria before reaching Turkey, routes that have always posing unacceptable risks for capital investment of tens of billions of dollars, and that became impossible with the disintegration of Syria and Iraq.

But Qatari gas reserves are too large and too inexpensive to produce to be left in the ground, and that is why Qatargas decided to develop them in two technically more demanding but ultimately rewarding ways, by building the world's largest GLT project (see Chapter 4 for details) and by acquiring the world's largest gas liquefaction capacity as well the most modern and largest fleet of supersized LNG tankers.

## 5.3    EVOLUTION OF LNG SHIPMENTS

Increasing volumes of the fuel are traded in the form LNG. Liquefaction entails cooling the gas down to $-162°C$ and reducing its volume to nearly 1/600th of the gaseous state: its density is 428 g/l (1 $m^3$ of $LNG = 0.43$ t) compared to 0.761 g/l for methane at ambient temperature.

Liquefaction is done in several independent units (trains) that are typically about 300 m long (Linde, 2013). Liquid gas is stored in super-insulated containers before it is transferred (via articulated pipes) to isothermal tanks on LNG tankers. Typical LNG contains 94.7% $CH_4$, 4.8% $C_2H_6$, and mere traces of higher alkanes and nitrogen, and its energy density is 53.6 MJ/kg (22.2 MJ/l, compared to 0.035–0.037 MJ/l for natural gas).

LNG is shipped in special tankers (Figure 5.5). The largest of these vessels (Q-Max ships of Qatar) carry 266,000 $m^3$ of gas (Qatargas, 2014a). *Mozah* (named after the wife of Qatar's emir) was the first ship of this class built by Samsung Heavy Industries in 2008, and it is now transporting LNG to the United Kingdom (Marine Traffic, 2014). Qatar now has 14 of these ships whose typical dimensions are a length of 345 m, a width of 53.8 m, and a draft of 12 m. With 22.2 $GJ/m^3$, a Q-Max ship's cargo amounts to almost 6 PJ of floating energy storage, enough to fuel 70,000 US homes for 1 year. Regasification takes place in seawater vaporizers. Most of the recently commissioned terminals have liquefaction capacities of 4–5 Mt/year, and most of the new receiving terminals rate between 3 and 5 Mt/year (IGU [International Gas Union], 2014).

Evolution of large-scale LNG industry is a perfect example of gradual advances whose progress has been determined not only by technical capabilities but also by willingness to pay initially very high costs for importing the gas from overseas. The long road to global LNG market began in 1852 when Thomas Joule and William Thompson (later Lord Kelvin) discovered that as a highly compressed air flowing through a

**Figure 5.5**    LNG tanker *Arctic Voyager*. © Corbis.

porous nozzle expands to the ambient pressure, it cools slightly (Almqvist, 2003). Sequential use of this incremental cooling was put into practice only by 1895 when Carl von Linde patented a reliable process of air liquefaction (Linde, 1916). Two decades later, Godfrey Cabot patented a design for shipping LNG by oceangoing vessels (Cabot, 1915), but as long as the United States remained the only major user of natural gas, there was no incentive to commercialize this possibility.

But there were important developments with LNG storage; the first small LNG project was completed in West Virginia in 1939 and a larger one in Cleveland in 1941, both to store energy-dense fuel for the periods of peak demand. Unfortunately, in 1944, a new larger, and poorly built, storage tank in Cleveland tanks failed, vaporized LNG ignited, and the resulting explosion killed 128 people. This accident, traced by a subsequent investigation to a poor tank design, came to haunt the industry as a proof that LNG storage, and by implication even more so the LNG transportation, is inherently too risky (USBM [US Bureau of Mines], 1946). Production of liquefied industrial gases expanded rapidly after WWII, but as North America only increased its global primacy as natural gas producers and as Europe and Japan embarked on imports of cheap Middle Eastern oil, there was no incentive to develop expensive LNG imports.

The first demonstration shipment of just 5,000 m³ of LNG (from Lake Charles, LA, to Canvey Island on the Thames) took place in January 1959. *Methane Pioneer* was not a true tanker but merely a converted Liberty-class vessel (built in 1945 as one of more than 2,700 leading WWII freighters), and it subsequently carried seven other LNG cargoes to Canvey Island before it was discarded in 1967. In 1961, the British Gas Board signed an agreement for long-term imports of LNG from Algeria's new Camel (Sonatrach) Plant in Arzew (three trains, low capacity of just 0.9 Mt/year) and the editorial in *The Gas World* (on November 11, 1961) called it "triumph of common sense". Deliveries of natural gas from Hassi R'Mel via Arzew to Canvey Island began in 1964, and two dedicated tankers (*Methane Princess* and *Methane Progress*) served the route, each carrying 27,400 m$^3$ (Corkhill, 1975). In 1972, Sonatrach added another LNG facility at Skikda, and both LNG plants were later expanded with additional trains.

Japan became the next importer LNG. In the absence of domestic hydrocarbon resources, it was motivated by desire to diversify its sources of imported energy (at that time only crude oil and coal) and reduce heavy air pollution in its large coastal cities. During the late 1960s, commercially the most viable opportunity was to bring LNG from Alaska, and in 1969, two tankers, *Polar Alaska* and *Arctic Tokyo* (each with capacity of about 70,000 m$^3$), began to peddle between Cook Inlet, Alaska, where Phillips Petroleum and Marathon Oil built a single-train LNG plant (annual capacity of 1.5 Mt) and Tokyo (Marine Exchange of Alaska, 2014). First imports from an Asian country began in 1972 when Brunei LNG was shipped to Ōsaka.

But as new LNG importers were added in the early 1970s (France from Arzew, Spain and Italy from a new Marsa El Brega plant Libya in 1970), the future of European LNG purchases began to look doubtful. Groningen, a giant Dutch natural gas field near Slochteren, was discovered on July 22, 1959, and as additional drilling uncovered its magnitude, it became clear that the find will not only end the age of coal in the Netherlands but that it will make the country a major exporter of natural gas to neighboring European states (Smil, 2010a). Concurrently, search for hydrocarbons in the North Sea had finally yielded both crude oil and natural gas: the first large gas find was in the West Sole Field in September 1965, 3 months later came the Viking Gas Field and more discoveries followed during the late 1960s.

As a result, the British contract for Algerian LNG imports was not renewed after its 1979 expiry, and the future of LNG imports to Europe became even more unlikely in 1984 with the completion of the first high-capacity pipeline (Urengoy–Uzhgorod) to bring Russian (at that time Soviet) gas from West Siberia. Concurrently, American LNG imports fell far short of initial expectations. The first US regasification terminal began to operate in November 1971 at one of the most unlikely locations, in Everett, Massachusetts, on the northern coast of the Mystic

River less than 4 km from Faneuil Hall, with tankers moving upstream past the city's downtown; moreover, the terminal is also less than 5 km from the NW–SE runway of Boston's Logan International Airport. This location focused attention on potential risks of LNG transportation and accidents during docking, loading, and offloading or due to earthquakes or a terrorist attack. Potential hazards arise from cryogenic temperatures of the transported gas, its dispersion, and flammability. Wilson (1973) was concerned about these hazards because he thought that their possibilities (particularly the explosion of spilled methane) bring to mind a comparison with potential nuclear reactor accidents. A few years later, Fay (1980) worried about new phenomena that could become evident in large-scale spills but cannot be observed in small-scale tests. And, not surprisingly, after 9/11, there were increased concerns about the LNG tankers as targets or tools of terrorist attacks (Havens, 2003).

But for half a century, the modern LNG industry has had a remarkable safety record demonstrated by Acton et al. (2013) in their study of 328 incidents that took place at peak shaving or LNG receiving facilities worldwide between 1965 and 2007. They found a very low frequency of incidents (0.24/site-year, most recently just 0.14/site-year), with about 75% of hydrocarbon releases being less than 100 kg/event and originating mostly from storage and send-out equipment. Just 7% of these incidents resulted in fire, explosion, or rapid phase transition, gravity of such events has decreased over time, and operation and maintenance shortcomings were the primary cause. The only category of increasing incidents is related to LNG trucks whose operation is often outside the control of LNG terminals.

Since the beginning of international LNG, there were no reports of damage outside peak shaving or receiving facilities, and hence, it is correct to conclude that imports, regasification, and distribution of LNG have not been a public hazard. As for the liquefaction plants, there was only one catastrophic explosion resulting in substantial loss of life and large economic loss. This most serious accident took place on January 19, 2004, in an LNG plant in Skikda on the Mediterranean coast of Algeria when a steam boiler (used to produce steam for turbines supplying power to compressors) exploded and triggered a massive secondary explosion of gas vapor cloud (Hydrocarbons Technology, 2014). Explosions and the subsequent fire (that took 8 hours to extinguish) killed 26 people, injured 74, destroyed three LNG trains and damaged another one, but left two trains, as well as all storage tanks, intact. Most remarkably, since 1964, there has been never any serious shipping accident resulting in loss of life and cargo or damage to port facilities as LNG tankers have crisscrossed the oceans delivering gas from every continent.

Two additional US terminals, at Cove Point in Maryland and Elba Island in Georgia, began to import LNG in 1978, but they were shut down just 2 years later. Lake Charles terminal in Louisiana opened in 1981 and was shut down the next year. LNG prospects dimmed further with the deregulation of US gas prices, with the 1985 collapse of high crude oil prices, and in 1986, no exported LNG came to the United States, but the shipments from Algeria to Lake Charles were resumed in 1988. Little had changed during the 1990s, the decade of relatively low and stable energy prices. This left Japan as the only major LNG buyer, and the country concluded new long-term import contracts in 1977 with the United Arab Emirates (Abu Dhabi) and Indonesia (LNG from Bontang field in eastern Kalimantan; a year later, shipments also began from the Arun plant) and in 1983 with Malaysia (Bintulu LNG plant), and by 1984, 75% of all traded LNG were sold to Japan. In 1991, LNG from Australia's North West Shelf was added to Japanese imports.

That began to change soon afterward as two East Asian countries, whose economic development was based on the Japanese path, became LNG importers: South Korea in 1986 and Taiwan in 1990 (both from Indonesia). And as in Japan's case, their purchases were in the form of long-term supply contracts, with LNG shipped from facilities with modest annual capacities and using relatively small tankers. During the late 1990s, two companies began to export LNG from the world's largest natural gas field: three Qatargas trains began shipping in November 1996 and two RasGas trains came on line in August 1999. Qatargas under the leadership of Faisal al-Suwaidi accepted Shell, ExxonMobil, and Total as full partners and went on to develop the world's largest LNG export capacity: 77 Mt/year is an energy equivalent of one quarter of Saudi Arabia's crude oil exports in 2012.

In 1999, Trinidad and Tobago's Atlantic LNG was the first company in Latin America to complete its liquefaction facility at Point Fortin on the southwest coast, and it began exporting to the United States (to Everett), while Nigeria's first two LNG trains began operating at Bonny Island in the Niger Delta. At the end of the twentieth century, the LNG industry was maturing, but it still had a long way to go before it could be seen as a key global player. There were 12 LNG-exporting countries, and a global fleet of about 100 tankers was used to export almost 140 Mt of the fuel to nine importing nations; that was about 20% of all natural gas exports but less than 6% of the global natural gas consumption (Castle, 2007).

Most LNG plants still had annual train capacities of just 1–2 Mt/year, the largest one could ship 3.5 Mt/year, and the volume of the largest LNG tankers had not changed for a quarter of the century: the record volume was 126,227 m$^3$ in 1975, and after surpassing 130,000 m$^3$ in 1978, the size of the largest tanker grew only marginally during the next 25 years, reaching 145,000 m$^3$ in 2003, a capacity stagnation mostly explained by the fact that Japan, the world's largest LNG importer, limited the maximum size of LNG tankers allowed to dock in its ports. Evolution of LNG exports during the twentieth century is thus an excellent example of gradual advances characteristic of global energy transitions (Smil, 2010a).

If we consider 5% of a market as a threshold of global importance, then it took LNG exports more than a century after the first commercial liquefaction to reach that threshold which came 50 years after the first 1959 test shipment from Louisiana to the United Kingdom. And while in the year 2000 LNG imports supplied significant portions of primary energy in Japan, South Korea, and Taiwan, they had only marginal roles in Europe and in the United States, and everywhere, they were dwarfed by imports of crude oil and refined oil products. But the first decade of a new century broke that long spell of slow advances, both in terms of expanding export base and fundamental technical improvements.

In the year 2000, Oman became LNG exporter and Greece and Puerto Rico became the latest new importers; by 2003, as the US domestic gas production kept declining, all of the country's regasification terminals were in operation for the first time since 1981, and in 2004, the first off-shore terminal was approved near Port Pelican, Texas. Ranks of LNG exporters were enlarged by Egypt (2004) and Norway and Equatorial Guinea (2007), while Portugal and Dominican Republic (2003), India (2004), and Mexico and China (2006) became the new importers. Most importantly, declining output and rising prices of the US natural gas pointed to an increasing need for large-scale LNG imports, demand that was to be satisfied not only by existing exporters but by the development of Russia's largest offshore field, Shtokman in the Barents Sea. And China's rising energy demand and extraordinarily high air pollution levels in the country's major cities indicated that the world's fastest growing economy will need many more LNG contracts than those already concluded with Australia and Indonesia.

In 2008, LNG shipments reached 25% of all natural gas, and the total annual export capacity had more than doubled during the first decade of the twenty-first century (from 100 Mt in 2000 to about 220 Mt in 2009) as maximum train sizes had more than doubled during the same time and as, after decades of stagnations, new class of large LNG tankers entered service, bringing obvious economic advantages but requiring solution of many technical challenges because of changes on loading, unloading, and vapor handling (Cho et al., 2005). Large aluminum spheres (Kvaerner–Moss shells) covered with insulation and placed inside steel tanks attached to the vessel's hull were the standard (and space-wasting and cumbersome) choice for more than three decades.

New membrane design uses thin stainless steel to make insulated tanks that fit the vessel's inner hull. Qatargas took advantage of this innovation by ordering 45 new large vessels belonging to Q-Flex (210,000 m$^3$) and to Q-Max (266,000 m$^3$) class (Qatargas, 2014b). This capacity growth meant that, after decades of no or slow growth, the volume of the largest LNG tankers had finally increased enough to compete with capacities that have been standard for large crude oil tankers since the late 1920s. But LNG tankers are still more expensive to build than oil tankers: 125,000–138,000 m$^3$ LNG tanker cost nearly twice as much (about $200 million) than a 250,000 dwt (VLCC) oil tanker that could be had for $120 million (IMO [International Maritime Organization], 2008). And because crude oil has 60% higher energy density than LNG (roughly 36 MJ/l vs. 22 MJ/l), large oil tankers carry four to five time as much energy per shipment.

But even as the LNG exports were rising and a strong consensus saw further rapid expansion of facilities, fleets, and capacities, two unexpected developments intervened and affected the pace of anticipated developments. The worst post-WWII economic downturn (2008–2010) reduced energy demand and affected future development of all major infrastructural projects, while hydraulic fracking (in 2005 a production method appreciated only by a relatively small number of experts and by 2010 one of the most widely publicize innovations in modern US history) became a major, and rising, contributor to America's natural gas (as well as crude oil) supply. The sudden emergence of fracking has led to three remarkable developments: to an impressive decline in North American gas prices, to an unprecedented decoupling of US oil and gas prices, and not just to cancellation of LNG imports plans but to plans for US LNG exports.

This is then where the global LNG industry stands half a century after its tentative British beginning (GLNGI [Global LNG Info], 2014; GIIGNL [International Group of Liquefied Gas Importers], 2014; IGU, 2014; BRS [Barry Rogliano Salles], 2014). By the middle of 2014, there were 27 LNG plants with 86 liquefaction trains in operation in 17 countries (Angola became the latest exporter in 2013), with 11 plants under construction and 22 in planning stages (Figure 5.6). Ras Laffan in Qatar was the world's largest LNG-exporting facility with annual capacity of 10 Mt (Qatargas, 2014a), and Qatar (with nearly a third of the total shipped volume) remained the leading exporter. Global trade reached 236.8 Mt in 2013 (slightly less than the record of 241.5 Mt in 2011) with just over 60% of all imported gas consumed in Asia–Pacific region, with Europe far behind (at about 33 Mt). About 77 Mt, or a third of all exported LNG, were traded on the spot and short-term market, with Nigeria and Qatar dominating the spot trade. Most of the contracts are still for long-term deliveries, with oil-linked prices at roughly \$12/million BTU in Europe and \$15 in Japan. LNG fleet consisted of 357 tankers with combined capacity of 54 Mm$^3$, and nearly 100 new vessels were on order.

Importance of LNG sales for the exporting countries is best illustrated by looking at the shares of LNG sale in their annual export earnings. Here are, in descending order, the ten top nations with the ordering based on the Standard International Trade Classification (export of item 3413: liquefied hydrocarbons) and calculated for the year 2010 by Hausmann et al. (2013) as a part of the Harvard–MIT economic complexity project: Qatar, 42%; Trinidad and Tobago, 34%; Algeria, 15%; Yemen, 15%; Oman, 14%; Egypt, 9.8%; Nigeria, 8.5%; Indonesia,

Figure 5.6    Australian Karratha LNG terminal for gas exports to Asia. © Corbis.

5.7%; Malaysia, 4.9%; and the United Arab Emirates, 4.4%. But several oil-exporting countries are more dependent on a single commodity than Qatar is on LNG: crude oil accounts for 89% of Azeri export earnings, and the other high shares are 86% for Nigeria, 76% for Saudi Arabia, and 67% for Kuwait.

In 2013, 29 countries were LNG importers and they had 104 regasification terminals. Three new importers were added in 2013: Israel, Singapore, and, remarkably, Malaysia, the world's second largest exporter (sales of 24.7 Mt in 2013, imports of 1.6 Mt). Regasification capacity reached a new record at 688 Mt/year which means that its utilization rate declined to just 34% in 2013, and floating regasification expanded to nine countries with a total capacity of more than 44 Mt. Japan, the largest importer, had the largest number of regasification plants (24), with Tōkyō Bay's Futtsu terminal, the world's largest, able to receive 19.95 Mt/year, most of it to power the world's second largest (4.5 GW) gas-fired electricity-generating plant (TEPCO [Tokyo Electric Power Company], 2013; Figure 5.7). Papua New Guinea has become Japan's latest source of LNG in May 2014, and the country's massive $19 billion project (taking gas from Southern Highlands and Western Provinces) will also supply China and Taiwan (PNG LNG, 2014).

LNG facilities have always made fairly small land claims, mostly just 30–120 ha. A liquefaction capacity of 3 Mt/year and an area of 80 ha

Figure 5.7    Futtsu LNG terminal. © Corbis.

would translate to a throughput power density of 6,400 W/m². The first three liquefaction trains at Ras Laffan (annual capacity of 10 Mt) claim 3.7 km² (Qatargas, 2014a), operating with a throughput power density of about 4,600 W/m². New modular designs are even more compact: 4.3 Mt Norwegian Snøhvit on Melkøya island near Hammerfest (4.3 Mt/year) occupies only about 70 ha and operates with processing power density of about 10,000 W/m² (Nilsen, 2012; see Fig. 8.4). Regasification plants have similar or even higher power densities. America's largest receiving LNG terminal, Cheniere's Sabine Pass in Louisiana, has a maximum throughput power density of about 8,300 W/m², Higashi–Niigata on the coast of the Sea of Japan (30 ha, 8.45 Mt/year) reaches nearly 48,000 W/m², and Tōkyō Bay's Futtsu terminal, the world's largest regasification plant (just 5 ha and 19.95 Mt/year), has a throughput power density of about 60,000 W/m².

# 6

# Diversification of Sources

Although the term nonconventional gas resources will be used for decades to come, extraction of some of those resources has become quite routine in the United States where nonconventional gas (shale gas, gas from tight sands, and coal-bed methane (CBM)) accounted for just 18% of the total output in 1990, over 60% by 2005, and 73% by 2012 (USEIA [US Energy Information Administration], 2014i). But the United States and, to a lesser extent, Canada are two great exception, and in most gas-producing countries, nonconventional gas sources are still only of marginal importance. These resources belong to four distinct formations: natural gas held tightly in shales (often also in sandstones), methane in coal beds, gas in high-pressure aquifers, and methane hydrates (clathrates) in the Arctic and in much larger undersea deposits.

The quest for difficult-to-extract natural gas is not new, and one of the methods that was not just proposed, but actually tried several times in the United States, is truly incredible (an adjective used with restraint). A more fitting label would be irrational, but its proponents (and there were many) in the US government were convinced that it would work: they thought that explosions of nuclear devices will increase the effective drainage volume and enable economic production of natural gas from tight sands. These experiments, part of the Plowshare Program, were characteristic of the era when many thought of nuclear energy as an effective enabler that could be deployed for many commercial tasks (US Congress, 1973; Kaufman, 2013).

The first test, Operation Gasbuggy in New Mexico on December 10, 1967, detonated a 29 kt device (Hiroshima bomb was about 12.5 kt of

*Natural Gas: Fuel for the 21st Century*, First Edition. Vaclav Smil.
© 2015 John Wiley & Sons, Ltd. Published 2015 by John Wiley & Sons, Ltd.

TNT equivalent) just 1,270 m below the surface, and a month after the detonation, a well was drilled into the resulting rubble chimney, and about 5.7 Mm³ of gas was withdrawn and flared at the site. In energy terms, about 120 TJ was released by a nuclear explosion in order to withdraw about 200 TJ of natural gas, a net energy return (1.7-fold) only if one disregards all the energy needed to produce the explosive nuclear device. The second test, Project Rulison in Colorado on September 10, 1969, exploded a 40 kt device, and during the third test, in 1973 in Rio Blanco County in Colorado, three 33 kt nuclear devices were detonated almost simultaneously in low-permeability sandstones 1.75–2 km below ground.

The fourth test, Wagon Wheel in Wyoming, never took place: it would have detonated five devices of 100 kt each at depths of 2.7–3.5 km, producing a rubble chimney more than 800 m tall and 300 m in diameter and releasing at least 35 times more energy than the energy of expected natural gas extraction (Noble, 1972). If successful, the US Atomic Energy Commission and El Paso had plans for full-scale production that would have included as many as 40–50 detonations every year! Reading this four decades later has only increased the sense on incredulity: how could these frequent detonations be ever justified in net energy terms, and how could regular detonation of powerful nuclear bombs underneath the grassland, fields, and forests of the American West be accepted by the public as routine means of producing gas used for heating and cooking? The experiments produced results far below expectations, and the project was abandoned.

But more promising techniques were under development when the nuclear route was abandoned. In his review of unconventional natural gas resources, Mankin (1983) noted that the process known as massive hydraulic fracturing (creation of vertical fractures extending from the wellbore into the reservoir) received a great deal of attention by the oil and gas industry (as well as federal research funding) but that it remained too expensive to recover gas from tight sands unless the producers receive subsidies or granted special incentives. He was no less pessimistic about natural gas from organic-rich shales, a challenge that cannot be easily overcome by hydraulic fracturing due to the problems with fracture propagation, recovery of fracturing fluid, and rapid decline of gas extraction.

His conclusion (published almost exactly 25 years before the large increases of oil and gas produced by hydraulic fracturing) is worth quoting in full:

In any case, it is reasonable to assume that any increased development of natural gas from Devonian shales will be confined to meet local or regional needs because of the low rates of production per well.

(Mankin, 1983, 35)

On the other hand, he was convinced that "the magnitude of the resource base and the long-range future for natural gas are factors that will lead ultimately to substantial development. In all likelihood, such development will not materialize until the early part of the 21st century" (Mankin, 1983, 42). That was a much better appraisal as large-scale development of shale gas released by horizontal drilling and hydraulic fractioning actually came in the second half of the very first decade of a new century.

Gas-bearing shales underlie large areas on all continents, but, so far, only the United States has developed this resource on a large scale (Maugeri, 2013; Zuckerman, 2013; Gold, 2014; Figure 6.1). Tracing the evolution of what is often called natural gas revolution, assessing the current state of hydraulic fracturing (commonly called fracking), and looking into the immediate future of shale gas will take up most of this chapter. The rest of it is devoted to advances and prospects for commercial recovery of other nonconventional gas resources: CBM, tight gas, and gas hydrates. China and the United States, the largest coal-mining nations, have also the largest resources of coal-bed gas, while tight gas

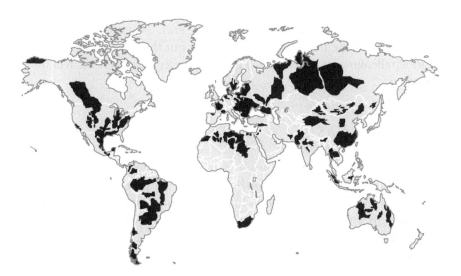

Figure 6.1   Global shale deposits.

(in formations with minimal permeability) is more widely distributed, and a large share of methane hydrates is offshore, deep under the seabed. But, before starting this systematic coverage with a more detailed focus on shale gas, I should first stress that the production of natural gas from nonconventional sources can be seen as a part of a larger progression of extracting mineral resources.

The sequence starts with tapping the resources that are easiest to find and easiest to produce with limited technical capabilities of early developmental stages; in the case of natural gas, the fuel itself was not initially even the object of search as it came associated with the eagerly sought-after crude oil and its production required only the separation of liquid and gaseous hydrocarbons. Later decades of deliberate quest for natural gas used the accumulating experience of oil exploration and extraction to bring large fields of nonassociated gas to the market; in many cases, that required not only higher capital investment in developing fields in extreme environments (Arctic, hot deserts, offshore) but also construction of long pipelines to bring the fuel to major consuming areas thousands of kilometers away.

As the most accessible and inexpensively produced reserves of mineral resources decline, the attention shifts to abundant deposits of lower quality or to remaining high-quality deposits whose recovery demands advanced techniques and rising production costs. Quality decline has been evident in the case of many common metals (as many extracted ores now have metal content an order of magnitude lower than in the early decades of the modern industrial era). Nonconventional gas resources yield the fuel of the same quality but their extraction is much more challenging and hence is highly dependent on relevant technical advances.

## 6.1  SHALE GAS

Assessments of shale gas resources and technically recoverable reserves and descriptions and reviews of basic operating procedures (involving horizontal drilling and high-volume hydraulic fracturing (HVHF)), regulatory frameworks, and common environmental impacts are readily available (USDOE [US Department of Energy], 2009; NETL, 2013; Speight, 2013). And yet writing about shale gas in concise and comprehensive way has become rather challenging, particularly when trying to give a balanced assessment of the US developments. Some controversial aspects of this development are only part of this challenge; the main hurdle is simply the cumulative amount of constantly increasing information, claims, and counterclaims brought forth by perhaps the most remarkable, and most consequential, expansion of resource extraction in modern history. Its progress has been so remarkable because it has been so sudden: for those in the know, it was not entirely unexpected, but even lifelong participants in the US oil and gas industry were surprised by the swift takeoff of the practice that led them to see it as revolutionary in nature.

In 2003, two experienced petroleum geologists published an assessment of North American natural gas supply that had a single perfunctory reference to shale gas (listing it among potential opportunities together with deep drilling, CBM, tight gas, and hydrates) and concluded that the days of cheap gas are history and that additional LNG facilities are needed for increased imports but because they cannot be built rapidly in significant numbers "rationing of available gas by the market mechanism of higher prices seems the only option" (Youngquist and Duncan, 2003, 229). In just 6 years, the United States, thanks to horizontal drilling and hydraulic fracturing, regained its global primacy in natural gas output, and many projections saw the country as a major LNG exporter even before 2020.

But this rapid shift was preceded by (mostly unappreciated) long technical gestation as individual components of new extraction technique made their gradual progress before they were combined in a mature and commercially rewarding process. During the early decades of drilling, it was difficult to keep straight course with deep wells, and deviated wellbores were fairly common. Deliberate directional drilling, able to exploit larger reservoir volume from a single well site, was the next step, and the first short horizontal well was drilled in Texas in 1929 (USEIA, 1993). Their high cost and absence of equipment most suitable to this task delayed the commercial adoption of horizontal drilling until the 1970s. More than 300 horizontal wells were drilled in North America during the 1980s, more than 3,000 during the next decade when the introduction of rotary steerable drilling motors introduced enabled to drill fairly easily even 90° bends and to follow undulating seams. Moreover, these motors can be mounted at the end of small-diameter (2.5–11.25 cm) coiled tubing, dispensing with rigid steel drill pipes while reaching a depth of up to 4 km.

Hydraulic fracturing had its distant precursor in oil well "shooting" done first during the early decades of the US oil expansion with nitroglycerine in order to increase to shatter the oil-bearing rocks and boost the liquid flow (Montgomery and Smith, 2010). Nonexplosive fluids were used for the first time during the 1930s to increase productivity through acid etching, but the first experiment with hydraulic fracking came only in 1947 in Kansas when Standard Oil of Indiana used pressurized fluid to create fractures and fluid-borne propping agents to hold them open in order to release previously unobtainable hydrocarbons.

This technique was patented in 1949 (Howard, 1949). The Halliburton Oil Well Cementing Company (Howco) became the exclusive licensee of the Hydrafrac process, and it used it for the first time on March 17, 1949 (Green, 2014). On that day, Howco performed two commercial fracturing treatments—in Stephens County, Oklahoma (costing $900), and in Archer County, Texas (costing $1000)—and in the first year, it treated 332 wells, boosting their production by an average of 75%. The technique was rapidly adopted by many producers and by the mid-1950s more than 3,000 wells were treated per month, and Montgomery and Smith (2010) estimate that between 1949 and 2010 the industry completed some 2.5 million fracturing treatments used for as many as 60% of all completed wells. The key goal is to create durable hydraulic fractions whose highly conductive flow paths would remain effective over decades of production (Vincent and Besler, 2013).

By 1990, both horizontal drilling and hydraulic fracturing were thus fully commercial procedures used overwhelmingly by oil industry to boost the productivity of crude oil extraction. How these techniques were adopted, adapted, and diffused by the US natural gas producers and deployed to extract natural gas from the country's extensive shale formations is complex, slowly unfolding story recounted by one of its protagonists (Bowker, 2003; Steward, 2007), described in books looking at the unfolding gas revolution (Levi, 2013; Zuckerman, 2013; Gold, 2014) and analyzed from historical and policy perspective (Wang and Krupnick, 2013).

Its beginnings go to the late 1970s, to government tax incentives to spur the development of nonconventional gas resources; its important components included several federally funded research programs (above all the Eastern Gas Shales Project at Morgantown Energy Research Center that examined the behavior of underground fracturing and brought advances in directional drilling) and continued advances in new mapping techniques and 3D imaging (that made it possible to pinpoint richest shale concentrations). Massive hydraulic fracturing was demonstrated for the first time in 1977 as a part of the Department of Energy project, the first horizontal well in Devonian shale was completed in 1986, but decisive breakthroughs toward commercial shale gas extraction came only during the 1990s.

George P. Mitchell, the founder of Mitchell Energy & Development, assembled a small team of experts and kept on defying persistent skeptics in his pioneering search for rewarding gas extraction from Texas Barnett shale. Mitchell's team took advantage of accumulating know-how coming from federally supported research, their first horizontal well in 1991 was subsidized by the Gas Research Institute, but it took large financial risks during its long pursuit of commercial gas from shales. In 1997, the company introduced a new, less expensive liquid formula containing less (and cheaper) sand and a gelling agent from guar bean.

## 6.1.1 American Shale Gas Extraction

Once the potential for economic extraction of the Barnett shale became evident, Devon Energy acquired Mitchell Energy & Development in January 2002 and began to combine its experience in horizontal drilling with water-based hydraulic fracturing: in 2002, there were 2,083 vertical wells in Barnett shale, and 5 years later, they were 8,960 wells, of which 55% (nearly 5,000) were horizontal (Brackett, 2008). By 2007, Barnett shale was still the dominant producer of the new resource, but its success led to the reassessment of other shale formations, and in 2009 came the news about the world's largest supergiant natural gas field—in Appalachia (Figure 6.2).

Most of this American region contains Devonian shales (they extend from south-central New York to southern Virginia and from eastern Ohio to northeastern Pennsylvania and are also underneath the eastern part of Lake Erie) that have been known for their kerogen content and hydrocarbon potential, but in 2002 the USGS put the technically recoverable volume of Marcellus shale gas in the Appalachian Basin at just 56.6 Gm$^3$ with additional 14 Mt of natural gas liquids (USGS [United States Geological Survey], 2011). Six years later, Engelder and Lash concluded that "the Marcellus Shale weighs in with more than 500 trillion cuft of gas in-place spread over a four state area. Continuous natural gas accumulations such as the Barnett Shale produce more than 10% of the gas in-place, which when applied to the Marcellus Shale, translates to a resource that will return 50 Tcf in time" (Engelder and Lash, 2008).

In standard units, that means almost 15 Tm$^3$ of gas in place and roughly 1.5 Tm$^3$ of ultimately recoverable volume, 25 times the 2002 USGS estimate. Because Engelder did not have enough public data to define a Marcellus decline curve, he used pro forma decline rate published in 2008 Chesapeake report in order to estimate ultimate recovery

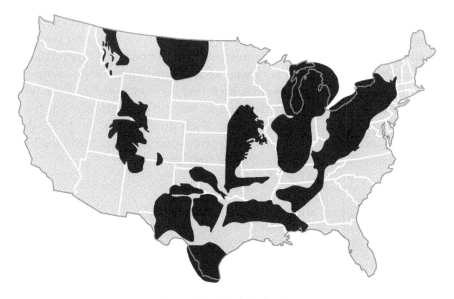

Figure 6.2   US shale basins.

from Marcellus; he also assumed that 70% of the sections in each county are accessible and that the wells have 80 acre spacing and ended up with 50% probability of 13.8 Tm³ of ultimate yield (Engelder, 2009). That was equal to just over 20 years of the country's annual natural gas consumption in 2009. Following Engelder and Lash, the USGS had also raised its estimate of recoverable Marcellus gas to an even higher total of 2.3 Tm³ (USGS, 2011), more than 40 times its 2002 total: I will return to these large discrepancies in estimating ultimately recoverable gas volumes in the book's last chapter when I will try to assess how far can natural gas go.

Following the Barnett breakthrough, other formations began to show rapid extraction increases. Fayetteville shale (part of Arkoma basin in Arkansas) began to take off in 2008, and by 2013, the gas output had roughly sextupled. Haynesville shale (underlying parts of southwestern Arkansas, northwest Louisiana, and East Texas) was the next one to take off, with average monthly extraction rising to more than 60% of Louisiana's gas output by the spring of 2010 after just 3 years of development (but its output declined by 27% between 2012 and 2013). Starting in 2010, Marcellus shale matched that rapid rise. Eagle Ford shale (extending in an arc from the Mexican border northeastward south of San Antonio), Woodford shale (underlying virtually the entire Oklahoma), and Bakken shale have been the other notable contributors.

Bakken shale (part of the Williston Basin centered on western North Dakota and extending to Manitoba, Saskatchewan, and Montana) has been North America's latest addition to large formations producing both oil and gas. In North Dakota, oil extraction (now higher than in Alaska and California and second only to Texas) by horizontal drilling and fracturing has been accompanied by so much natural gas that (in the absence of adequate pipeline capacity) large volumes of it had to be flared: by the end of 2011, more than one-third of all gas produced in North Dakota was flared or not marketed (USEIA, 2011).

Data on new-well gas production per rig show substantial differences among major shale basins, both in daily rates and in trends between 2007 and 2014 (USEIA, 2014j). In the Permian Basin, average performance has actually declined by half to about 3,000 cuft/day; in Eagle Ford shale, it has remained steady at around 1,200 cuft/day; in Niobrara, it has been up and down; Bakken has seen steady but slow rise to less than 600 cuft/day; and Haynesville and Marcellus have been the star performers: in the former formation, the mean had quintupled between 2007 and 2014 to more than 5,000 cuft, while Marcellus has seen an order-of-magnitude gain, from less than 600 to more than 6,000 cuft/day. In aggregate, in the year 2000, shale gas supplied just 1.6% of the total US natural gas production; in 2005, it was 4.1%; in 2010, it was 23.1%; and in 2013, it reached 40%. This unprecedented ascendance of a new energy resource has engendered both exaggerated hopes and, this being America of hyperbolic claims that need not be based on any realities, also near apocalyptic warnings.

The development of shale hydrocarbons has resulted in local and regional impacts well known from previous cases of resource booms as well in some consequences specific to the industry. Sudden influx of new workers (dominated by single males) creates shortages of suitable housing, drives increases in real estate prices, and overtaxes planning and zoning capacities of smaller settlements that are overwhelmed by new permitting requests. Relatively high earnings in the industry increase the cost of business for other sectors, create labor shortages in low-paid service establishments, and prevent economic diversification because intensive resource development lowers the chance of attracting concurrently other investments; the latter reality only worsens longer-term economic prospects once the initial expansion stage is over, and depopulation rate may be higher than in places with no comparable resource development.

Environmental impacts include some well-known phenomena as well as some new annoyances and damages. As with any kind of hydrocarbon exploration, drilling, and production, there will be local surface land disturbances, destruction of plant cover, and habitat fragmentation. Good news is that shale gas drilling is now often done from space-saving multiwell pad sites taking up 1.6–2.0 ha during the fracturing phase but only 0.4–0.6 ha after site restoration (NYSDEC [New York State Department of Environmental Conservation], 2009; Figure 6.3). Certainly, the most obvious change specific to HVHF is an enormous increase in truck traffic required to deliver fracking fluids. Inevitably, this leads to road congestion on two- or single-lane rural roads, road deterioration (to the point of impassability for unpaved roads), rising road maintenance costs, reduced quality of life (as inhabitants of formerly isolated houses near a rural road suddenly see hundreds of large truck going down their formerly quiet lane), and, most unfortunately, more fatal accidents (Begos and Fahey, 2014).

While air and noise pollution caused by heavy tanker truck traffic and high-pressure fracturing is only temporarily disruptive, concerns about potential degradation of groundwater quality are much longer lasting, and, not surprisingly, they have attracted a great deal of public attention and protests (Kharaka et al., 2013). Antifracking movement had coalesced first around the concerns about fracking liquids contaminating local aquifers and about drinking water laden with pollutants

Figure 6.3   Shale gas drilling site in Pennsylvania. © Corbis.

that could be set on fire when it comes out of a faucet (this became an iconic image of fracking gone awry). Only later came concerns about longer-term effects of air pollution from escaping gases and the increased potential for localized earthquakes. As with all similar movements, antifracking activists form a heterogeneous group whose members range from professional environmental protesters opposed to virtually any resource developments to people genuinely concerned about fracking wells few hundred meters upwind from their backyards, schools, or playfields.

As expected, the cause has attracted its share of celebrities, including not only such veterans of assorted causes as Yoko Ono—whose measured assessment is that "Fracking kills. And it doesn't just kill us, it kills the land, nature and eventually the whole world"—and Robert Redford but also Alec Baldwin and an Iron Chef Mario Batali (Begos and Peltz, 2013). By September 2013, the antifracturing sentiment reached even to Dallas where the city council rescinded leases granted previously to Trinity East Energy and where at the public hearings the opponents outnumbered those who favored the development, citing increased cancer risk (disproved by actual statistics) at a Colorado fracking site (Ginsberg, 2013).

And Bamberger and Oswald (2014) believe that large-scale fracking has many public health implications, particularly for animals, children, and oil and gas workers. They also compared many people affected by the industry to victims of rape because they are powerless and at the complete mercy of forces beyond their control. Abrams (2014) amplified their claims by publishing her interview with the authors under the title "Fracking's untold health threat: How toxic contamination is destroying lives." Some of these harmful exposures have led to successful lawsuits. On April 22, 2014, a family in Dallas County won $3 million in the first Texas case verdict that found a fracking operator (Aruba Petroleum) guilty of fouling the plaintiff's ranch property, their home, and quality of life and sickening them and their pets and livestock (Matthews & Associates, 2014).

Opposition has been also considerable in New York State, where the movement began with seeking a permanent ban on any drilling in the city's watershed and where New Yorkers Against Fracking seek to make the entire state off-limits because they consider the activity to be "in the same category as smoking.... The only way to make smoking safe is to not smoke" (Navarro, 2013). Statewide ban was announced in 2014. I will review water requirements of HVHF and the complex evidence regarding drinking water contamination and wastewater disposal in the next chapter.

## 6.1.2   Shales outside the United States

While the disputes about the consequences of fracking continue in the United States, environmental concerns have been a key reason why several European countries—including France, the Netherlands, Czech Republic, and Bulgaria—enacted preemptive bans (or at least temporary moratoria) on shale fracking. To be sure, there are other factors that militate against US-like European embrace of natural gas. Perhaps none as important as the fact that, unlike in the United States where mineral rights come with the land ownership, the continent's landowners do not own the rights to develop resources under their land, a reality that does not give them any incentive to get involved in potentially very disruptive extractive activities. Other important differences are a much more flexible access to pipeline capacity and, of course, much lower population density in many of America's major shale regions and hence a reduced probability of NIMBY conflicts.

This leaves plenty of other countries with the opportunity to develop a new energy resource because hydrocarbon-bearing shales (with organic content of 2% and more) are among the world's most commonly encountered formations. Besides the United States, other large nations with extensive shale formations include Canada, Brazil, Argentina, Russia (in Western and Central Siberia), Algeria, South Africa, Pakistan, China, and Australia. The first worldwide resource assessment concentrated on 32 countries and only on the regions known to contain within their shale basins the higher-quality prospective areas (hence generally biasing the totals in conservative manner); it included two key judgments regarding the likelihood of potential gas flow for commercial development and an expectation of the extent to which such prospective areas within each shale gas basin and formation will be developed in the foreseeable future (Kuuskraa et al., 2011).

This assessment ended with the total of 623 Tm$^3$ of gas in place and with 163 Tm$^3$ of technically recoverable resources. China's recoverable volume was put slightly ahead of the United States, with Argentina, Mexico, and South Africa constituting the top five nations. Uncertainty of these estimates is best illustrated by a reassessment of recoverable gas issued by the same organization (Advanced Resources International) just 2 years later (Kuuskraa, 2013). Norway's recoverable resources went from 2.3 Tm$^3$ to zero due to disappointing results from three wells drilled by Royal Dutch Shell in Sweden's Alum shale, a similar nearby formation. In a similar case of a near-total downgrading in 2014, the USEIA cut its estimate of recoverable oil from California's Monterey shale by 96%

compared to its initial claim made in 2011 (Reuters, 2014): clearly, more large reappraisals, and indeed eliminations of technically recoverable shale hydrocarbons, must be expected as we learn more about specific formations.

In 2013, Kuuskraa cut Libya's estimated shale gas volume by nearly 60%, total for France was reduced by 25%, and those for South Africa and Mexico went down by 20% (Kuuskraa, 2013). At the same time, newly assessed basins in Ukraine and Algeria tripled their 2011 totals and added nearly 50% to the Canadian total (as already noted mostly in the Montney formation), and these additions resulted in an overall 35% increase of technically recoverable shale gas worldwide to about $220\,Tm^3$ or about $7.7\,ZJ$. That is nearly 20% more than the BP's (2014a) latest global estimate of conventional natural gas resource at the end of 2013 ($185.7\,Tm^3$) and nearly four times the total energy ($1.94\,ZJ$ in 335 billion barrels) in recoverable shale oil. But this, too, is only a temporary value, and we will get a more realistic understanding of what is really technically recoverable only after every major shale basin will be tested by a significant number of exploratory wells.

Reduction of shale gas recovery for China was minor (by <10%) but large enough to put the United States in the first place—but the Chinese situation is a near-perfect example of complexities and specificities that must be considered when assessing realistic chances of shale gas development rather than producing generic theoretical estimates based on the extent of shale-bearing basins and their organic content (Chang and Strahl, 2012; Tollefson, 2013). Although China had originally plans for extracting at least $60\,Gm^3$ (and up to $80\,Gm^3$) of shale gas by 2020, it has had virtually no experience in applying and adjusting the two key constituent techniques of horizontal drilling and hydraulic fracturing and limited number of people with requisite expertise. No less importantly, Chinese shales lie deeper than in major US formations, are more scattered, have more fractures and faults, and are inadequately mapped, and hence, they will be almost certainly more costly to extract. Not surprisingly, in August 2014, Chinese production goal for shale gas in 2020 was cut by half to $30\,Gm^3$ (Shale Gas International, 2014).

Water availability is another major consideration: Sichuan Basin has plenty of precipitation, but it is also China's most populous province with huge water demand for irrigation, industries, and megacities, while western shale basins (Tarim, Junggar) are arid and already experience serious shortages of water (Marsters, 2013). And institutional and entrepreneurial barriers (including state dominance of oil and gas industry, absence of experienced, risk-taking small enterprises, willingness to deal with failure, an almost inevitable outcome of pioneering new production methods in difficult conditions) should not be forgotten. And China will not be the only nation where shale gas development is unlikely to replicate the post-2007 American experience. While blanket bans on hydraulic fracturing (à la France) make little sense, many expectations will be disappointed and many opportunities that might become profitable after periods of trials, errors, and adjustments will be missed because of early failures. In 2013, Talisman Energy of Calgary and Marathon Oil of Houston withdrew from shale gas exploration in Poland, a country that embraced the development of this new resource put by Kuuskraa et al. (2011) at $5.2 \, Tm^3$ of recoverable gas in 2011 but reduced by the Polish Geological Institute to just $800 \, Gm^3$ a year later. Development of Mexico's large shale gas resources has been off to a very slow start, with fewer than 25 wells drilled by mid-2014. And the extraction of Argentina's giant Vaca Muerta shale gas deposit is complicated by the country's notoriously precarious fiscal situation and uncertainty about its domestic energy policy.

## 6.2   CBM AND TIGHT GAS

America's success in shale gas extraction has diverted attention from other kinds of nonconventional source of natural gas, and although these resources may not be developed with a similar speed and at a comparable extent, they will become increasingly more important contributors to gas production. Kuuskraa (2004) emphasized their importance by pointing out that 8 out of 12 of the largest US gas fields are nonconventional: the largest two, Blanco and Basin in New Mexico, are a combination of tight gas and CBM; numbers five, six, eight, and nine (in Texas, Colorado, and Wyoming) contain gas in tight formations; and number four (Wyodak in Wyoming) is a giant coal bed.

A great deal of methane gets generated during coal formation, and its content in coal seams is generally higher in deeper and older layers: more pressurized deposits hold more gas, with typical content of just $0.02 \, m^3/t$ of coal at depth of $100 \, m$, $1 \, m^3/t$ at $500 \, m$, and about $3.7 \, m^3/t$ at $1,000 \, m$ (IEA [International Energy Agency], 2005). But the gas is very tightly bound within the fuel's complex structure that has virtually no permeability. But the solid fuel has many cleats (natural opening-mode fractures) that account for most of coal's permeability and most of the porosity of coal-bed gas reservoirs (Laubach et al., 1998). Because of its large internal surface area, coal stores six to seven times more gas per unit volume than gas-bearing rocks in conventional hydrocarbon reservoirs—but even the latest and relatively most effective drilling and extraction methods are able to recover only a small fraction of the resource in place.

Methane associated with coal makes the news when an accidental explosion of the gas in improperly ventilated underground mines causes deaths of miners. To use a proper technical term, the gas responsible for these explosions is coal mine methane (CMM) that has been released from coal seams or from collapses of surrounding rock following underground fuel extraction. Proper treatment will dilute the gas below the explosive range (preferably to <1% by volume) and it is removed by large ventilation systems. The gas can be also recovered in order to be converted to $CO_2$ by oxidation or to be used in lean-burn gas turbines, and more than 200 of these processes in 14 countries help to avoid annually nearly $4 \, Gm^3$ of $CH_4$ emissions or prevent wasteful flaring (WCA [World Coal Association], 2014).

In contrast, CBM is the gas recovered from unmined coal seams, no matter if they will or will not be eventually extracted. Gas recovery begins with removing coal seam water in order to reduce pressure and release the adsorbed gas; afterward, vertical and horizontal wells drilled through unmined seams are used to recover the gas, and hydraulic fracturing may be used in some seams to release more of the gas from coal seams. Production rate of CBM wells thus has three distinct phases: the first one dominated by water, the second one by rising gas flow and declining water flow, and the third one with both gas and water flows in decline (Garbutt, 2004). The gas has usually a high concentration of $CH_4$ (>93%), and hence it could be used directly as a replacement for conventional natural gas and transported in existing pipelines.

Large-scale commercial extraction of CBM began in the United States during the 1980s; in 2012, it contributed 5% of gross natural gas withdrawals; and its volume was equal to about 15% of shale gas extraction in that year (USEIA, 2014k). In absolute terms, that was nearly 25% below the peak reached in 2008, a consequence of rising extraction of shale gas. Similarly, large CBM resources in Russia remain undeveloped because of the abundance of Siberian natural gas. Obviously, all nations with extensive coal deposits—the United States, Canada, Australia, Russia, China, India, and South Africa—have significant resources of CBM, but so far, only Australia began their large-scale recovery: Queensland's high-quality bituminous coal seams now supply about a third of all gas in the eastern part of the country, adding to its huge resources of conventional offshore gas (Geoscience Australia, 2012).

An older US estimate indicated about $20\,Tm^3$ of coal-bed gas in place, with about $3\,Tm^3$ economically recoverable (Rice, 1997), but Kuuskraa (2004) concluded that technically recoverable reserves (with 2002 capabilities) are only $820\,Gm^3$. IEA (2005) put the global CBM resources at $143\,Tm^3$. Kuuskraa (2004) estimated global resources in place at $85–222\,Tm^3$ (IEA, 2005). With a typical recovery of about 20% of all gas in place, this means recoverable volume of $17–44\,Tm^3$, compared to $185\,Tm^3$ of proved conventional natural gas reserves in 2012 (BP [British Petroleum], 2014a). The largest extraction potential is in the United States, Canada, Russia, and China, the greatest expansion of CBM recovery expected in China where clean fuel alternatives are limited. There is also considerable potential for recovering methane from abandoned coal mines in old mining regions in Europe (the United Kingdom, Germany, Czech Republic), in the United States, and in China. Existing old ventilating shaft and new wells are used to extract the gas.

## 6.2.1   Tight Gas

Tight gas is methane stored in rocks—mostly in sandstones but also in siltstones, limestones, dolomites, and chalk—with very low *in situ* matrix permeability, almost always less than $0.1\,mD$; recall that most conventional natural gas reservoirs have permeability between $0.01$ and $0.5\,D$, that is $10\text{-}500\,mD$. Tightness of these formations is a consequence of their age: while conventional natural gas comes mostly from younger Tertiary basins, tight gas originates mostly from Paleozoic formations with very low permeability due to prolonged compaction, cementing, and recrystallization. China's Sulige field discovered in the Ordos Basin in 1999 is an excellent example of challenges facing the harnessing of natural gas in tight formations (Total, 2007; Yang et al., 2008; CNPC [China National Petroleum Corporation], 2014; Figure 6.4). Its permeability is between $0.02$ and $2\,mD$; gas-bearing layers created by an extensive braided river system 250 million years ago are just a few meters thick and are $3.2–3.5\,km$ below the surface. Yet Sulige is the world's largest sandstone trap whose reserves may be as much as $2\,Tm^3$, and its expensive extraction, began in 2006, should reach $5\,Gm^3/year$.

Because the recovery depends on penetrating as much formation volume as possible, tight gas production depends on drilling the largest (economically justifiable) number of directional and horizontal drilling, preferably from a central drill pad. Additional production stimulation is achieved by hydraulic fracturing as well as by introducing acids to

Figure 6.4 Sulige field in China. © Corbis.

dissolve basic formation substances (limestone, dolomite, calcite) and hence to enhance permeability of tight sediments. Removing water from deeper wells also increases their productivity.

Global resources of gas in tight formations may be as large as $510\,Tm^3$, but existing recovery techniques yield only 6–10% of gas in place, reducing the technically recoverable reserves to no more than $50\,Tm^3$. Estimates of the US tight gas in place have been as high as $26\,Tm^3$ in 1980, Kuuskraa (2004) put technically recoverable total at more than $8\,Tm^3$, and USEIA estimates the total at about $8.8\,Tm^3$, considerably larger than even the highest appraisals of CBM and almost a fifth of the country's total gas reserves. Only a few companies outside the United States have extensive experience with producing gas from tight formations; notable commercial developments have been in Algeria (Timimoun), Venezuela (Yucal-Placer), Argentina (Aguada Pichana), and, as already noted, China (Sulige).

American producers have a fairly long history of extracting tight gas; they began to do so during the 1960s in the San Juan Basin in New Mexico where they had pioneered the relevant recovery techniques. US tight formations are widely distributed—in the Appalachian Basin, Denver Basin (mainly in Colorado and Wyoming), West Gulf Coast Basin (in Texas, Louisiana, and Mississippi), Permian Basin (in West Texas), Uinta Basin (in Utah), and Greater Green and Wind River Basins of Wyoming (USEIA, 2010)—and the United States is the only country with relatively large-scale natural gas production from tight sands, recently about 170 Gm³/year or about 25% of all gas produced.

## 6.3 METHANE HYDRATES

These remarkable energy stores (also known as clathrates or clathrate hydrates) were created as the gas from methanogenic archaea feeding on organic sediments escaped and became trapped inside lattice cages formed by molecules of frozen water (Kvenvolden, 1993; Makogon, Holditch, and Makogon, 2007; Figure 6.5). Fully saturated gas hydrates (at 2.6 MPa) contain one molecule of $CH_4$ for every 5.75 molecules of $H_2O$, and this means than one m³ of methane hydrate holds up to 164 m³ of methane. Because 80% of hydrate volume is taken up by water and 20% by methane, one cubic meter of hydrate contains 164 m³ of gas. One m³ of water can tie up to 207 m³ $CH_4$ as it forms 1.26 m³ of solid hydrate; without gas, freezing 1 m³ of water expands its volume to just 1.09 m³ of ice.

Specific density of hydrates (depending on composition, pressure, and temperature) ranges from 0.8 to 1.2 g/cm³, and the gas composition is dominated by methane, ranging from as slow as 66% to as high as

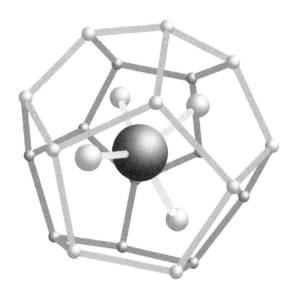

Figure 6.5    Methane hydrate cage.

99.7% $CH_4$ (Taylor, 2002). Hydrates can be stable under an enormous range of pressures and temperatures (from just 20 nPa to 2 GPa, which is 17 orders of magnitude, and from 70 to 350 K). Hydrates form when water and gas are subjected to low temperature and elevated pressure, an all too common occurrence in natural gas pipelines where hydrate plugs can damage equipment and where precautions must be taken to limit their formation. The solid plugs were first ascribed to the freezing of condensed water, but Hammerschmidt (1934) proved them to be a hydrate of the transported gas. Interest in hydrates as a potential source of commercial energy began in 1963 with the drilling of the Markhinskaya well drilled in Yakutiya (Central Siberia): the presence of a hydrate layer led Makogon (1965) to conclude that gas hydrate accumulations might be found wherever conditions are right to create cooled layers.

In 1965 and 1966, Makogon's laboratory experiments demonstrated the formation hydrates, and by 1970, the presence of hydrates was confirmed in the West Siberian Messoyakha field. The field began to produce natural gas from a reservoir situated beneath the gas hydrate layer, but its output was soon overshadowed by such supergiant West Siberian fields as Urengoy and Medvezhye. But some geologists concluded that the depressurization of Messoyakha reservoir also led to depressurization and gas dissociation from the overlying hydrate formation, making it the only instance where conventional production methods would yield gas from hydrates. But a reexamination of accumulated evidence by Collett and Ginsburg (1998) suggested that gas hydrates may not have contributed to gas production in the Messoyakha field.

Distribution of hydrates is obviously restricted by temperature of surrounding sediments. The two natural environments favoring the formation of hydrates are sediments in the Arctic (only about 3% of the total resource base) and sediments beneath the ocean floor. Some hydrates appear as small nodules of just 5 cm in diameter, others form continuous pure layers several meters thick. More than 200 gas hydrate deposits have been identified, mostly in waters off the coasts of Americas and East Asia and in the western Pacific (Figure 6.6). Some hydrate deposits in the North American Arctic are only about 100 m below the surface. Among the studied ocean hydrate deposits those off California are just 600 m below the sea bottom and underneath 600 m of water, while those in the deepest parts of Japan's Nankai Trough are in water depths of around 4,700 m and 4,800 m below the sea bottom, and off Guatemala both depths are still another 1,000 m greater (Makogon, Holditch, and Makogon, 2007).

Figure 6.6    Methane hydrate global deposits.

Not surprisingly, estimates of the total volume of methane held in hydrates are even more uncertain than the shifting assessments of shale gas or coal-bed gas. But these differences are only of secondary importance because even very conservative estimates indicate the enormity of resources in place. Dillon et al. (1992) put the mass of organic carbon in methane hydrates at 10 Tt or roughly twice as much as carbon stored in all conventional fossil fuels. Lowrie and Max (1999) estimated that the volume of gas beneath the seabed of the US coastal waters may be as much as 1,000 times the volume of all US conventional gas reserves.

The range for Canadian resources is between 43.6 and 809 Tm$^3$ (Center for Energy, 2014). Makogon, in many publications, uses a grand global total of 15 Pm$^3$ (which is roughly 15 Pt) of potential reserves, and Kvenvolden (1988) went as high as 20 Pm$^3$, which is almost exactly 100 times the total volume of conventional global gas reserves in 2013. And a recent assessment of methane reservoirs beneath Antarctica showed that some 21 Tt of organic carbon is buried in deep organic sediment underlying the continent and that the sub-Antarctic hydrate inventory of 131–728 Tm$^3$ could be of the same order of magnitude as the latest estimates for the Arctic permafrost (Wadham et al., 2012).

Several options might lead to commercial recovery of hydrates: simply exposing their formations to lower pressure created by well; removing water and other gases from wells drilled into hydrate formations;

releasing the gas by injecting steam or hot water in order to disassociate the gas hydrate; and injecting $CO_2$ into hydrate reservoir in order to displace a substantial share of methane. The first test of decompression and heating was done in Canada in 2001. Gas hydrates have been encountered for decades in many wells drilled in Canada's Mackenzie Delta, and one of these sites was chosen by an international (Canada, Japan, the United States, Germany, India) project at the Mallik Bay Gas Hydrate Research Well conducted in late 2001 and 2002 (Dallimore et al., 2002).

Three wells were drilled (one for production, two flanking wells for monitoring), and pressure reduction was tested in six zones as a 17 m layer with high gas hydrate saturation was heated by hot water for several days. This showed that commercial production from land-based hydrates might be eventually feasible, and during winter of 2008 a 6-day test (139 hours) at the Mallik site was able to maintain flows of 2,000–4,000 m³/day with cumulative gas production of about 13,000 m³ (Yamamoto and Dallimore, 2008).

Shale gas development in North America had soon displaced all other gas-related news, but hydrates were back in March 2013 with news from Japan where hydrate research program began in 1995. Its first important finding was the discovery of extensive methane hydrate deposits with high $CH_4$ concentrations (gas filling 60–90% of the pore space) in sand reservoirs off the coast of Japan. This was a welcome departure from a common occurrence of hydrates in unconsolidated muds where they fill only 10% of the pore space (Boswell, 2009).

The Japanese test was conducted by Japan Oil, Gas and Metals National Corporation, and research ship was used to drill a 270 m deep well to reach a 60 m thick hydrate reservoir 1 km below the ocean bottom and about 80 km off Atsumi Peninsula (the east coast of Japan) at Daini Atsumi Knoll in the eastern Nankai Trough (JOGMEC, Japan Oil, Gas and Metals Corporation, 2013). Pump was used to reduce the formation's pressure from 13.5 to 4.5 MPa, letting the gas rise and flow to a platform on the ship where it was flared starting on March 12, 2013. Although the designers tried to prevent the intake of sand by installing two sifting devices, the pump clogged on the sixth day of the trial, but up to that point the gas flow averaged 20,000 m³/day, an order of magnitude more than obtained by the Canadian depressurization test, and the total yield was about 120,000 m³.

This was an encouraging experiment as significant volumes of gas were produced for nearly a week, but no commercial breakthroughs are imminent. Not surprisingly, North American interest (the US hydrate research program began in 1982) was diverted by shale gas development, but besides Japan, there are relatively small research programs in South Korea (since 1999), China (since 2004), and India (since 1996). Advances of these efforts and other hydrate news can be followed in *Fire in the Ice*, a periodical publication by the US National Energy Technology Laboratory (NETL, 2014).

# 7

# Natural Gas in Energy Transitions

Development of modern economies has been characterized by a number of universal transitions. They included changes of sectoral contributions (as the dominance of wealth creation shifted from agriculture to industrial production and then to services), labor participation (the end of child labor, entry of women into labor force, postponed age of the first employment due to longer periods of education), and capital intensity (from artisanal manufacturing requiring often no or minimal capital inputs to very expensive megaprojects undertaken by national governments). The most fundamental shift has been the evolution of modern energy supply as the urbanizing and industrializing societies had first moved away from traditional biofuels (wood, charcoal, in rural areas also crop residues) to coal and then to refined liquid fuels and natural gas while concurrently generating more primary (hydro, nuclear, and now also wind and solar) electricity.

Wood, and to a lesser degree crop residues (mainly staple grain straws and stalks), dominated millennia of premodern and the three centuries of early modern (1500–1800) energy supply. Its replacement by coal began first (already before 1600) in England, but in the continental Europe and in the United States, it took place, fairly rapidly, only during the latter half of the nineteenth century. When leaving traditional biofuels aside, coal delivered about 96% of global total primary energy supply (TPES) by 1900, with crude oil at about 3% and natural gas (overwhelmingly due to the US consumption) a mere 1%

*Natural Gas: Fuel for the 21st Century*, First Edition. Vaclav Smil.
© 2015 John Wiley & Sons, Ltd. Published 2015 by John Wiley & Sons, Ltd.

(Smil, 2010a). Subsequent transition from coal to oil was slowed down by the two world wars and by the global economic crisis of the 1930s, but by 1950, coal's share of global TPES declined to just below two-thirds, while oil rose to roughly 30% and natural gas was still the distant third at 10%. The next two decades saw the emergence of new, postwar economies based on inexpensive crude oil: the United States remained its largest producer, but the output in Saudi Arabia and the USSR was catching up fast.

By 1973, when OPEC's actions led to the quintupling of global oil price, oil was the world's leading fossil fuel, accounting for 48% of the global TPES, with coal a receding second at 27% and natural gas rising to about 18%. Oil's further ascent was checked by OPEC's continuing oligopolic manipulations (the second round of price rises had more than tripled the 1974 rate by 1981) that spurred the quest for higher conversion efficiency of refined fuels and for their replacement by other energies. Oil's absolute consumption continued to grow, but by the year 2000, the fuel's share fell to 38% of the global TPES and as gas supplied about 24% of the global demand. During the first dozen years of the twenty-first century, oil's share fell further to 33%, and while the global extraction of natural gas rose by nearly 40%, its share of the TPES remained at about 24% due to a rapidly rising combustion of coal (thanks largely to expanded extraction in China and India) that pushed the fuel's share to roughly 30% of the world's TPES and had temporarily reversed the long-term energy transition from coal to hydrocarbons (in 2000, they supplied about 62% of the total and in 2012 only 57%).

When looking at the pace of energy transitions, I found some remarkable similarities between the global shift from traditional biofuels to coal and the subsequent substitution of coal by crude oil (Smil, 2010a; Figure 7.1). Coal reached 5% of the global TPES (a share I use as the marker of nonmarginal contribution) by about 1840, and its share was at 10% by 1855, 20% by 1870, 25% by 1875, 33% by 1885, 40% by 1895, and 50% by 1900. That spacing of milestones—15, 30, 35, 45, 55, and 60 years—is very similar to the sequence of identical milestones reached by crude oil as it was displacing coal and biofuels after it reached 5% of the global TPES (15, 35, 40, 50, and 60 years to reach 40%; oil will never supply 50% of TPES).

In comparison, the worldwide transition to natural gas has proceeded at a slower pace: the fuel reached 5% of the global TPES around 1930, 10% 20 years later, and 20% only 45 years later. And depending on the choice of energy conversions (above all the different ways of converting primary electricity to a common energy denominator), it has either

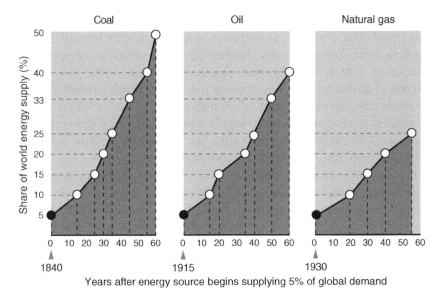

Figure 7.1 Global fuel transitions.

reached 25% after some 70 years, or it has yet to reach that level (BP has it at 23.0% in 2012)—while coal claimed 25% of the fuel market 35 years after it reached 5% and oil did so after 40 years. Obvious absence of any acceleration in successive transitions is significant: moving from coal to oil has been no faster than moving from traditional biofuels to coal—and substituting coal and oil by natural gas has been measurably slower than the two preceding shifts.

Scale of the requisite transitions is the main reason why natural gas shares of the TPES have been slower to rise: replicating a relative rise needs much more energy in a growing system. In the year 2000, the global TPES was about five times larger than in 1950 and roughly 10 times the size of the supply in 1900, and hence (even in the absence of any resource constraint), it has become progressively harder for a new source to claim a significant share of the overall demand. As a result, raising oil's share from 5 to 25% of the global TPES by 1945 called for about 1.6 times more energy than completing the same shift for coal by 1875—but going from 5 to 25% of natural gas required nearly eight times more energy than accomplishing the identical coal-to-oil shift.

Infrastructural challenges are the other obvious explanation as moving solids and liquids is much easier, especially across and among continents, than transporting natural gas. This slower pace means that those virtu-ally always unreliable long-term consumption forecasts have been even more amiss as far as the market penetration of the natural gas has been concerned. But even this slower-than-anticipated shift toward natural gas has been most welcome because the fuel offers a less polluting alternative that is also the best practical means to lower the carbon burden of global energy consumption that still depends overwhelmingly on fossil fuels. Replacing fuel oil by natural gas to heat a house or coal by natural gas to generate electricity is a much easier task than replacing coal or oil by carbon-free renewables because it can be done economically and with assured availability and reliability.

In contrast, even when affordable, solar or wind electricity generation cannot guarantee constant availability—unless backed up by other forms of on-demand generation (gas turbines being one of the best choices!) or by requisitely large storage—but the storage option is practical only on a relatively small scale as we have no means (save for relatively rare and inevitably energy-losing pumped hydro facilities) of storing electricity to meet demand on the scale of hundreds of MW or a few GW (power flows required by today's large cities). Consequently, all carbon-free renewable alternatives will have limited impact for the foreseeable future.

There is still considerable unexploited hydroenergy potential in Asia, Africa, and Latin America, but there are only few remaining prospects for large-scale hydroelectricity generation in North America and Europe. New renewables above all solar photovoltaics and wind have been capturing higher shares of national electricity generation in some countries (Germany being the leader among large affluent economies), but their market penetration has been proceeding at a relatively slow pace when measured as a share of the TPES: in 1990, 88% of the world's TPES came from fossil fuels; in 2012, the share was 87% (Smil, 2014). In addition, nuclear generation has stalled, or is declining, in most affluent countries; its future is made even more uncertain by the Fukushima disaster and continuing problems with the site's cleanup. Natural gas is thus the best option for near- and midterm decarbonization of the global TPES.

As already described (in Chapter 5), recent advances in LNG transportation have finally made a global gas market (akin to the long-established oil trade) an increasingly appealing economic option and have greatly improved the prospects for accelerated market penetration, but an even more important factor to determine the eventual outcome will be the extent and the pace with which natural gas can penetrate a key energy market in modern economies, that of transportation fuels. At the same time, we cannot assume that much higher dependence on natural gas will have only positive environmental impacts, and that is why I will assess all of the fuel's environmental consequences in the closing segment of this chapter.

## 7.1 FUEL SUBSTITUTIONS AND DECARBONIZATION OF ENERGY SUPPLY

Inevitably, the universal process of energy transitions—including the sequence of fuel substitutions from wood to coal to oil to natural gas and higher reliance on primary electricity—has displayed many nation-specific variations. Some countries had never went through a coal stage as they moved from wooden age to economies based on crude oil, other nations (most notably China) remain still highly dependent on coal, and yet others rely on exceptionally high shares of hydroelectricity. But different transition paths have eventually the same important outcome as societies benefit from higher efficiency of final energy conversions, from less pollution associated with combustion from reduced intensity of carbon emissions. Gradual decarbonization of energy supply is an especially desirable trend as it helps to moderate the human interference in the global carbon cycle and slow down the rise atmospheric $CO_2$.

Wood, the dominant energy source of the premodern and early modern (1500–1800) world, is mostly cellulose, hemicellulose, and lignin. Overall carbon content of these biopolymers averages about 50%, with a relatively narrow range of 46–55% (Lamlom and Savidge, 2003; Cornwell et al., 2009). Wood contains only about 5% of hydrogen, as does bituminous coal whose carbon content is around 65%. Atomic H:C ratio of wood is thus about 1.4 and of bituminous coal typically around 1.0, and hence, it would seem that the transition to coal did not result in any decarbonization of fuel use. But because a large share of wood's hydrogen atoms is never oxidized (due to hydroxyl radicals that escape in early stages of combustion), the effective H:C ratio of wood is typically less than 0.5, and a shift to coal results in producing lower $CO_2$ emissions per unit of fuel energy, and in practice, the reduction has been even greater due to higher combustion efficiencies of better designed stoves or boilers burning coal.

Liquid fuels derived from crude oil average 86% carbon and 13% hydrogen, and both gasoline and kerosene, with atomic H:C ratios of 1.8 (nearly twice as high as bituminous coal), resulted in a major shift toward decarbonization—and the gain is even greater when burning methane whose atomic H:C ratio is obviously 4.0. Specific carbon emissions thus decline from about 30 kg/GJ of wood to as little as 25 kg/ GJ of excellent bituminous coal and 20 kg/GJ of refined liquid fuels, and burning of natural gas will release only 15.3 kg C/GJ (IPCC [Intergovernmental Panel on Climate Change], 2006). Obviously, global energy transitions (whose major features were quantified in this chapter's opening section) have resulted in a persistent decline of specific carbon emissions (Figure 7.2).

When expressed in kg C/GJ of the global TPES, the rate declined from nearly 28 in 1900 to just below 25 in 1950 and to just over 19 in 2010, roughly a 30% decrease (Smil, 2013a). Among the affluent countries, the drop has been largest in France (due to its nuclear commitment, not because of natural gas), the US decline was nearly 40% (due to gas and nuclear), but China's only about 25% (due to continuing high reliance on coal). When this worldwide decarbonization is quantified in terms of H:C ratio of fossil fuels, its global mean rises from 1.0 in 1900 to 1.6 by 1950 and 1.8 by 1980 and 1.9 by the year 2000. Subsequent slight reversal (to almost 1.8 by 2012) was caused by China's massive addition of new coal extraction capacities.

This means that decarbonization of the global TPES has been proceeding at a slower rate than has been expected. Most notably, during the mid-1990s, Ausubel foresaw the global H:C mean of 3.0 in 2010

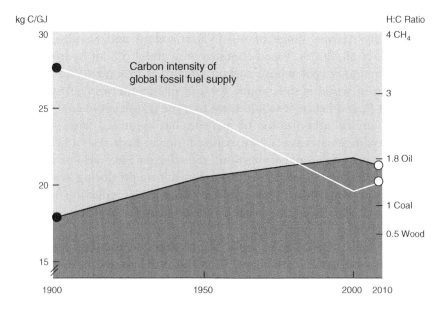

Figure 7.2    Decarbonization of global energy supply.

(the actual rate was just 1.83) and the global $CH_4$ economy (H:C ratio of 4.0) shortly after 2030, and more recently, he still expected that $CH_4$ will supply 70% of the global TPES soon after 2030 (Ausubel, 2003). That is not going to happen, but how far the substitutions will go remains uncertain, and I will assess the best evidence in the closing chapter. In any case, declines in the specific carbon intensity have yet to be translated into absolute decreases of global $CO_2$ emissions (CDIAC, 2014).

The verdict is clear: rising consumption of natural gas has been a key cause of decarbonizing the global TPES, but the recent growth of gas supply could neither prevent further growth of carbon emissions nor slow down their growth to such an extent that the world would avoid going above 450 ppm of $CO_2$ in decades to come. Much has been expected of natural gas, but there is only so much it can deliver in the world of rising energy demand. And, obviously, a much higher share of carbon-free energies (be it solar, wind, or other renewable modes of electricity generation and nuclear fission) will be needed to carry eventual decarbonization beyond the carbon limits inherent even in a pure $CH_4$ system.

As expected, national experiences have shown a wide range of outcomes, from rapid market penetration in smaller countries with abundant domestic resource of natural gas to still very low shares of the TPES in some of the world's largest energy consumers heavily dependent on imports. There is no better example of a rapidly executed shift to natural gas than the Dutch exploitation of the supergiant Groningen field. As already noted, the field was discovered in July 1959, and in that year, domestic coal dominated the country's primary energy supply (about 55%) followed by imported crude oil (about 43%), and natural gas supplied less than 2% of the total supply (UNO [United Nations Organization], 1976).

Groningen field began producing in December 1963, and as its output began to rise, the Dutch government decided in December 1965 (when the natural gas share stood at 5% of primary energy supply) to end all coal extraction in old Limburg fields (going back to the sixteenth century) in no more than 10 years. This deliberate social dislocation (Limburg mines employed 45,000 miners with 30,000 directly related jobs) was eased by giving the state mining company a 40% share in the gas development and helping it to reinvent itself as a producer of chemical goods and later into nutrition, pharmaceutics, and materials (DSM, 2014). By 1971, Groningen gas was providing half of the country's energy demand, and by 1975, it had stabilized at just short of 50%, while coal (mainly for coking) sank to less than 3%. And because during the early 1970s it was widely believed that nuclear energy will dominate the supply in the long term, it was also decided to maximize exports to the neighboring countries, and they had quadrupled between 1970 and 1980 to more than $40\,Gm^3$ by 1980. The shift to natural gas was swiftly accomplished as its share in the nation's TPES rose from less than 5% in 1965 to 33% by 1971. That was an unprecedented speed of substitution: after reaching the 5% mark, it took only less than 6 years to reach 33% (in comparison, the United States took 50 years to go from 5 to 25%, while the USSR needed 20 years to go from 20 to 40%). Extraction (further spurred by the belief that exports should be maximized before nuclear energy, at that time a very promising mode of future energy supply, would weaken the demand) peaked in 1977 at $82.3\,Gm^3$. The supply share during the 1980s and the 1990s leveled off at about 40% as the output became regulated (nuclear electricity did not take over!) in order to extend Groningen's lifespan. Meanwhile other, smaller, onshore and offshore fields began to augment the field's production, and by 1990, they supplied more than half of the Dutch gas as Groningen's output declined from the peak of more than $80\,Gm^3$ to less than $30\,Gm^3$ in the year 2000—but natural gas was still at 40% of the Dutch TPES in the year 2000 and at nearly 37% in 2012 (BP [British Petroleum], 2014a).

Domestic, and international, consequences of the Dutch natural gas extraction have been truly transformative. All space heating (household, commercial, institutional) was converted to Groningen gas, as was nearly all industrial processing and (a very important consideration for a small economy that is the world's second largest exporter of agricultural products by value) all heating of the country's extensive greenhouses. They add up to the world's largest area of heated cultivation (occupying more than 10,000 ha or roughly five times as much as land devoted to potatoes) and produce vegetables (in terms of value dominated by red and yellow peppers), fruits, and flowers, about 40% for export (TNO [Toegepast Natuurwetenschappelijk Onderzoek], 2008).

Moreover, part of $CO_2$ from the clean-burning natural gas is not emitted outside, but it is used to enrich the atmosphere inside greenhouses (to about 1,000 ppm compared to the ambient level of 400 ppm) in order to increase the rate of growth and greenhouse productivity (Hicklenton, 1988; NGMA [National Greenhouse Manufacturers Association], 2014). Exports of the Dutch gas helped the neighboring EU countries to reduce air pollution by displacing coal and fuel oil, to increase combustion efficiency of household heating and industrial processing, and to lessen the dependence on Russian natural gas exports or on even more expensive

LNG from the Arab countries—while earning the Netherlands on the order of $10 billion a year (Trading Economics, 2014).

Comparison of the British and the Dutch experience illustrates how national specificities affect the pace of fuel substitutions. Both countries profited from major natural gas discoveries (the first British North Sea gas discoveries were made in 1965), but the Dutch accomplished in six years (going from 5 to 33%) what the United Kingdom took 25 years to do: the North Sea gas began to supply more than 5% of the British TPES in 1971, and it reached 33% only after 26 years in 1997. Then its contribution peaked at 39% in the year 2000 and declined to less than 35% by 2012. Explanations of this disparity are obvious. When the transition began, the British energy was nearly four times larger than the Dutch requirement; the country's electricity generation was dominated by large coal-fired plants that could not be suddenly shut down; increasing share of electricity was coming from nuclear generation pioneered by British reactor designs; developing offshore resources was more challenging than putting Groningen on stream; and undersea pipelines and longer land lines had to be built to bring the gas to the British market.

Inevitably, the transitions have been even slower when the gas had to be imported from overseas: as already noted, Japan began its LNG imports in 1969; they reached 5% of the country's TPES in 1979, 14% by the year 2000, and (following the temporary closure of all nuclear power plants) 22% by 2012. That is a remarkably high share for which the country had to pay by incurring its first prolonged trade deficits since the early 1980s (Trading Economics, 2014). And it is almost certain that the imports will have to increase in the future because in the near term even a vigorous promotion of renewable conversion would not make for the losses of electricity-generating capacity due to the post-Fukushima (March 2011) shut down of Japan's nuclear power plants.

Recent surge in LNG imports put Japan far ahead of China whose natural gas extraction contributed only about 2% of the country's TPES until the end of the twentieth century. Subsequent tripling of domestic natural gas extraction and the rise of both pipeline (from Turkmenistan) and LNG imports (mostly from Qatar, Australia, Indonesia, and Malaysia) pushed the gas share to roughly 7.5% by the year 2012. That is the lowest relative contribution among the world's major economies as even India was slightly ahead in 2012, with nearly 9%. And although large LNG projects (now underway or in planning stages) will further increase the import capacity, the relative contribution of natural gas will continue to rise only slowly as long as the country's coal expansion will continue: it is now targeted to go to 4.8 Gt by 2020, nearly 40% up from 3.5 Gt in 2013 (Xinhua, 2013).

America's early transition from coal and oil to natural gas was relatively slow for two obvious reasons: the United States was the first major economy to pioneer the shift, and in its early years (as already noted) the pace was limited by technical capabilities (above all the absence of long-distance high-throughput pipelines); and the magnitude of the country's TPES precluded any rapid fuel substitutions. Natural gas reached 5% of the US primary energy production already in 1924 and 10% in 1935. The economic crisis and WW II slowed down the pace a bit, but 20% mark was reached in 1951 and 25% in 1957. A temporary peak just short of 34% came in the early 1970s, and declining production lowered it to about 25% by 1990. Relative gas contribution stagnated at that level for more than a decade until shale gas extraction pushed the share up to 31% by the year 2012 (Figure 7.3). This increase has been accompanied by a higher share of gas used for electricity generation: volume of gas used for thermal electricity generation had doubled between 1983 and 2013 as the share of all gas-derived electricity rose to nearly 30% (USEIA [US Energy Information Administration], 2014e).

Canada's rich natural endowment has made it the country with the lowest specific carbon emissions among the affluent economies. Until the late 1990s, carbon-free hydroelectricity supplied a higher share of its

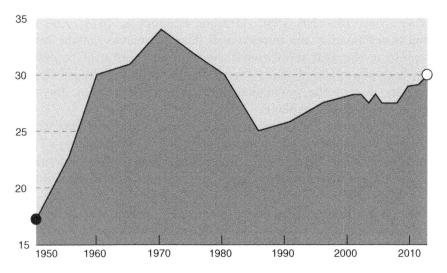

**Figure 7.3**    US gas share in primary energy production.

TPES than natural gas; as a result, the share of gas in primary energy consumption has been rising only slowly, taking 50 years to double to nearly 30% by the year 2012 (but for most of that time, at least 30% and up to almost 50% of Canada's annual natural gas production were exported to the United States). And so it is Mexico that has the highest natural gas share of the TPES in North America, at 40% in 2012, second only to crude oil and 30% above both the US and Canadian shares.

The post-1950 Soviet exploitation of natural gas deposits was initially delayed by large increases in oil production that pushed the fuel's share of the TPES from 16% in 1950 to the peak of 35–37% between 1974 and 1983. But between 1970 and 1990, rapid development of supergiant Siberian fields had more than quadrupled the Soviet gas production; by 1983, the USSR surpassed the US output and even its new large export commitments to supply its Communist satellites; and the EU were no impediment to a steadily rising gas share from less than 10% of the TPES in 1960 to more than 20% in 1970, 32% in 1980, and 41% in 1990. And while many performance figures have plummeted in the post-Soviet Russia, natural gas gained relatively greater importance as its supply share rose to about 52% by the year 2000 and to 54% in 2012.

## 7.2   METHANE IN TRANSPORTATION

This section should open by a reminder (see Chapter 3) that natural gas has had for decades a critical role in its own transportation, as efficient and reliable gas turbines power compressors that propel gas through pipelines. But the gas has, so far, been only a marginal source of energy for moving goods and people, and this minor contribution to the transportation sector presents a major impediment to the fuel's rise to a significantly higher share in the global primary energy supply. Transportation claims about 20% of the world's TPES and refined oil products supply 93% of that total and natural gas less than 4% (IEA [International Energy Agency], 2013). Transportation's sectoral share in the United States is higher (28%), but the shares of liquid fuels and natural gas are very similar at 93% and less than 3% (USEIA, 2014e). Of course, the reason for these marginal contributions is obvious: while natural gas is the best fuel for space heating and industrial processing, its low specific density and low volumetric energy density make it an inferior transportation fuel compared to superior densities of liquid fuels refined from crude oils.

This reality excludes it from ever being a fuel for commercial aviation: even after liquefaction, its volumetric energy density of 21.4 GJ/m is only about 60% that of volumetric energy density of jet fuel (kerosene)—and well-insulated tanks are needed to keep the fuel at -162°C. But even when solving the technical challenge of carrying the cryogenic fuels, the planes would need more than twice as much LNG to cover the same distance, the need that would inevitably limit their passenger and cargo capacity and make flying quite expensive on that account alone. But liquefaction and portable cryogenic storage do not present any insurmountable technical problems for using natural gas in heavy-duty land and water transportation where additional mass necessary to carry the fuel can be readily accommodated.

## 7.2.1   LNG

Indeed, LNG's benefits in heavy-duty applications—above all in shipping and trucking—make it an excellent choice for markets that are now served by high-efficiency diesel engines. Costs aside, the fuel offers three fundamental advantages when compared to diesel fuel: lower generation of $CO_2$ per unit of useful energy, lower emissions of sulfur and nitrogen oxides, and noise reduction. But LNG is also a fairly cost-competitive fuel choice, particularly for newly built ships and new truck fleets. LNG tankers are obviously the most obvious candidates, and they have used the fuel without mishaps for the past 40 years. They were originally propelled by steam turbines with steam raised by burning the naturally boiled-off gas, but soon, as in every form of mass cargo shipping, highly efficient and inexpensively operating diesel engines became dominant.

The two basic types of these prime movers are known as diesel low speed and reliquefaction (DLR) and dual-fuel diesel electric (DFDE). DLR uses low-speed diesels for propulsion and four to five auxiliary machines that power cargo pumps, supply onboard electricity, and also reliquefy any boiled-off gas and pump it back to the ship's tanks, with all DLR machines burning heavy fuel oil. DFDE propulsion consists of four to five identical engines to power the alternators, and the generated electricity is used for propulsion, pumps, compressors, and all auxiliary systems and accommodation needs. The DFDE's great advantage is that it can burn any mixture of naturally boiled-off gas and liquid fuel (heavy fuel oil or marine diesel). Naturally boiled-off gas amounts to 0.11–0.15% of the cargo load, enough to produce about 20 MW of kinetic energy for a tanker carrying 150,000 m³ of the gas. That could supply roughly two-thirds of the total demand when sailing fully loaded at 20 knots (37 km/h), and it would be far higher than 1.5 MW needed when a ship waits to enter a terminal or 7.5 MW needed when discharging its cargo (Smil, 2010a).

As the world's fleet of LNG ships expands, so will the use of LNG in dual-fuel engines of new tankers, but there is a much larger potential as LNG-powered propulsion can be adopted by other vessels: LNG tankers aside, in 2013, more than 30 ships were powered by LNG, and another 30 were to enter operation within 2 years, but that total is still only about 0.1% of all ocean-going cargo vessels. A key factor that will govern the rate of future adoption of (or conversion to) LNG will be the implementation of stricter air pollution rules, globally and for existing emission control areas (ECA) in North America and Western Europe and their eventual ECA extension to Asian coastal waters. The global limit on sulfur in fuel is 3.5%, while the existing ECAs—the Baltic Sea and the North Sea, 200 nautical miles zone along the Canadian and US coastal waters (including southern Alaska and Hawaii) and the US Caribbean— got 1% sulfur limit since 2010, and this demand is met by new low sulfur fuels, 380 CST and 180 CST (Kaur, 2010).

But standards should become much tougher, down to a mere 0.1% S in the ECAs beginning in January 2015 and globally to just 0.5% in 2020, with latter target contingent on the outcome of 2018 feasibility study (Adamchak and Adede, 2013). That assessment may show that there is not enough clean liquid fuel on the market to meet the target— but the compliance would be easy with LNG whose sulfur content is just 0.004%. In any case, several companies are planning to create LNG supply for a variety of ships, including ferries, barges, tugs, cruisers, and support vessels for offshore hydrocarbon production, laying of undersea pipelines, and construction of wind turbines. Semolinos (2013) anticipated that by 2020 LNG will capture about 5% of the overall marine fuel market, or about 3% of all LNG use, and that these shares will rise, respectively, to 10 and 5% by 2030.

Analysis of world ship traffic data by Burel, Taccani, and Zuliani (2013) shows that roll-on/roll-off vessels and small- and mid-size tankers (10,000–60,000 DWT) spend most of their time sailing in the ECAs and they would be the most suitable candidates for LNG-powered propulsion. Cleaner and quieter operation of LNG-powered engines would be particularly welcome on Europe's heavily frequented inland waterways. Since 2013, two Dutch-built LNG-powered barges deliver liquid fuels along the Rhine between the Netherlands and Switzerland, and their spark-ignition engines are up to 50% less noisy compared to conventional heavy diesels (Shell, 2013).

Using LNG as fuel for trucks is the best practical alternative to oil-derived fuels in vehicular transportation and the one that also has several advantages (Linde, 2014). As with LNG in shipping, the combustion of LNG has lower emissions of local pollutants (particulate matter, $NO_x$, and $SO_x$) than diesel fuel while also generating less $CO_2$ per unit of useful energy. Moreover, for high fuel-use fleets, LNG offers lower life-cycle cost than diesel. Truck drivers appreciate that the fuel is nontoxic and noncorrosive, that refueling is as simple and as fast as filling with diesel or gasoline, that the range is good (up to 1,000 km), and that LNG-powered engines are quieter than conventional diesel motors at normal speeds (an advantage to be appreciated by people living along heavily traveled roads; at high speeds and on steeper roads, the noise is similar) and operate with much lower vibration. For countries with plenty of domestic natural gas but with limited or no production of crude oil, LNG trucking obviates imports of refined liquid fuels.

LNG trucking is a particularly appealing option for new, rapidly expanding fleets, with China and India being the two best examples, but benefits would be also great on congested European and US roads. The European Commission has a demonstration project for LNG-powered heavy-duty vehicles that includes participation of major truck makers (IVECO, Volvo) and eventual creation of four blue corridors (Atlantic, Mediterranean, South–North, West–East) with refueling stations for medium and long trucking distances (Hubert and Ragetly, 2013). Similarly, in 2012, Shell revealed plans to build a corridor of LNG fueling station in Alberta to serve heavy-duty truck fleets delivering materials to the province's northern oil sands operations.

Supply chain for road LNG uses (as well as for some coastal traffic and inland waterways) would start with small-scale liquefaction, either on fixed or movable (then preferably modular) systems with annual capacities up to 1 Mt (compared to medium-sized LNG facilities of 2–3 Mt/year and large plants above 4 Mt/year). Shell has developed the moveable modular liquefaction system that can deliver LNG on a smaller scale, enough to meet local and regional transport requirement but requiring much lower capital expenditure than the standard large liquefaction facilities. Trailer trucks (or LNG bunker vessels) would distribute the fuel to LNG refueling stations (Figure 7.4).

But the progress has been slow and uneven. Early LNG adopters have experienced mixed results. Some have found that higher maintenance cost and lower-than-expected operating efficiency have greatly extended

Figure 7.4    LNG filling station. © Corbis.

their expected payback periods; others had higher-than-expected energy needs for diesel fuel required for compression ignitions and for in-cab methane detectors. In 2013, by far the largest North American LNG truck order was by the UPS for 700 vehicles, with the second highest order in the United States for just 36 trucks (Raven) and the largest Canadian order (by Bison transport) for only 15 trucks (Truck News, 2014)—in a market that sells about 250,000 units a year to an expanding fleet of more than 3.5 million units. In 2014, FedEx announced plans to fuel 30% of its fleet by LNG before 2025, Procter & Gamble aims at 20% share within 2 years, but in 2014 Shell pulled back on its LNG expansion plans in Alberta.

China had about 50,000 LNG-powered trucks in 2012 with plans for nearly 250,000 in 2015, and that would be still less than 5% of the country's heavy-duty truck fleet (Hong, 2013). But the Chinese–owned Blu LNG scaled down its plans for the US expansion after it had built only half the number of LNG refueling stations it planned to have in place by the end of 2013 (Groom, 2014). Moreover, the newest 12l gas-powered engine (a joint venture by Cummins and Westport) is best suited for smaller loads on flatter routes, and most of the orders have been for its compressed natural gas (CNG)-fueled configuration rather than for LNG.

## 7.2.2  CNG

At about 20 MPa, CNG is nearly 130 times as dense as ambient methane (volumetric density of 128.2 g/l compared to 0.761 g/l), but it has a much lower density than LNG (428 g/l). Much as other gas-fueled machines, CNG-powered vehicles reduce emissions of urban air pollutants: compared to diesels, CNG delivery trucks generate 95% less particulate matter, 50% less $NO_x$, and 75% less CO, while $CO_2$ reductions for buses are between 13 and 23% (Werpy et al., 2010). Marbek (2010) study found the following reductions of greenhouse gas emissions compared to diesel or gasoline engines: 23% for heavy-duty LNG-fueled and 19% for medium-duty CNG-fueled trucks and 23% for passenger cars running on CNG. In addition, unlike with LNG, refueling with CNG does not require any mask and gloves, and CNG-powered trucks are easier to maintain than LNG vehicles.

But road transportation powered by CNG faces more obstacles than does LNG for trucking, and it would not be enough just to remove or to lower one or two of them before CNG-fueled vehicles could find a wider acceptance in countries with high degree of automobilization and highly developed car market that offers highly competitive combinations of vehicles, fuels, and engines, with hybrids, plug-in hybrids, purely electric vehicles, and new clean diesels in the mix. The gap in model choice is enormous: North American market now offers more than 250 car models sold by more than 40 manufacturers, but only four US companies are making natural gas engines (conversions of diesels), and in 2013, American Honda was the only car manufacturer offering CNG vehicle (Civic GX costing about $5,000 more than gasoline-fueled Civic EX), while GM and Chrysler began to sell CNG-fueled pickups (with $11,000 premium) in 2012 (Fraas, Harrington, and Morgenstern, 2013).

Model choice is also limited in a few low-income countries that have promoted CNG vehicles and set up their domestic manufacturing: Toyota and Suzuki in Pakistan are the best example, but most Brazilian CNG vehicles are, as is the case in the United States, after-market conversions. Conversions (done by certified outfitters) include pressurized storage tanks (which take up useful space and increase vehicle's tare), pressure reducer (to the level of the engine's fuel-management system), shut-off valve, and fuel injectors. Conversion are expensive: Marbek (2010) study estimated additional capital investment of $8,000 for cars, about $50,000 for buses, and $90,000 for long-haul trucks. Obviously, these costs reduce the advantage of cheaper fuel, lengthening the payback period for a typical passenger car (driven less than 20,000 km/year) to 6–8 year compared to about 3 years for long-haul trucks. Higher maintenance requirements (also due to the need for periodic high-pressure tank testing) add to the lifetime ownership cost.

Refueling remains a challenge in most countries: in December 2012, Pakistan had some 3,300 stations and China nearly 2,800—but the US total was just 1,120 (compared to 160,000 serving gasoline); France, 149; and Canada, just 47 (IANGV [International Association of Natural Gas Vehicles], 2014). Moreover, a large share of these stations is not open to public. Putting in place a requisite refueling infrastructure poses a chicken-and-egg dilemma that is shared by electric vehicles: what come first, investment in widely accessible refueling infrastructure or a large vehicle base?

The cost of an additional refueling infrastructure would not be low. Gallagher (2013) estimated that the investment needed for CNG fuel dispensing in the US would be $100–200 billion (but only $10–20 billion for widely available LNG fleet refueling). And a limited range of CNG vehicles (about 150 km on a full tank) is another inconvenience.

All of these barriers explain why centrally fueled urban and suburban vehicle fleets operating within a limited area are the best candidates for fueling by CNG, either in bifuel mode (two separate fueling systems) or as vehicles powered only by gas. Model choice is largely irrelevant while suitable engines are readily available, conversion costs are repaid more rapidly for vehicles driven at least 50,000 km/year, and centralized refueling is no problem. That is why most CNG vehicles now operating in modern economies belong to high fuel-use urban fleets (Figure 7.5). Almost 20% of America's urban buses run on CNG, and in 1998, the Supreme Court of India ordered complete switch to CNG, later enforced by heavy fines to operators of diesel buses who refused to accept new vehicles (Marbek, 2010). Other common CNG-powered vehicles include garbage and delivery trucks, taxis, and shuttles.

Given these realities, it is not surprising that CNG-powered transportation accounts for only a tiny fraction of vehicular traffic: in 2012, the global share was 1.28%, ranging from negligible shares in affluent economies (the United States, 0.05%; Japan, 0.06%; France, 0.03%) to 77% in Armenia, 65% in Pakistan, and 37% in Bolivia (IANGV, 2014). In absolute terms, the two leaders were Iran and Pakistan (3 and 2.9 million vehicles, respectively) followed by Argentina (2.1 million), Brazil (1.75 million), and China (1.6 million). The US total was about 128,000 (overwhelmingly fleet) vehicles (of 253.7 million) served by about 1,100 refueling stations.

Figure 7.5    CNG bus in New Delhi. © Corbis.

Clearly, the choice has the greatest appeal in lower-income countries where CNG helps to ease specific challenges: shortage of oil refining capacity in Iran, need to reduce expensive oil imports in Pakistan and Bolivia, and high air pollution in China's cities. But even there, the switch may not ease the greenhouse gas emissions. Alvarez et al. (2012) concluded that replacing gasoline cars by CNG light-duty cars would bring no reductions of radiative forcing for 80 years and that for heavy-duty diesel vehicles there would be no gain even after more than 100 years. But CNG might offer an alternative to shipping smaller volumes of stranded gas. Smaller carrying capacity of CNG-fueled tankers proposed by Japan's Kawasaki Kisen Kaisha (equivalent of just 12,000 t of LNG) would be outweighed by less expensive fuel preparation (compression to 20 MPa vs. liquefaction to −162°C) and cheaper vessel construction ("K" Line, 2014).

## 7.3   NATURAL GAS AND THE ENVIRONMENT

Why should we worry about methane in the environment? The gas has been a part of the biosphere for billions of years; it is produced by a variety of natural processes and then is effectively removed from the atmosphere by oxidation. Moreover, as stressed in the opening chapter, methane generates less $CO_2$ per unit of useful energy than does the burning of any kind of wood, coal, crude oil, or refined oil products, and hence, it ranks as the best carbon energy source in the age concerned about global warming and rising carbon emissions. But this great advantage comes with an undesirable property: methane is the only fossil fuel that is also a greenhouse gas and, indeed, a more potent one than $CO_2$. Every molecule of $CH_4$ absorbs more outgoing long-wave radiation than does a molecule of $CO_2$, and approximately one–fifth of the increase in anthropogenic radiative forcing during the past 250 years has been due to rising atmospheric methane concentrations.

## 7.3.1   Methane Emissions from Gas Industry

By the end of the twentieth century, the global atmospheric burden of the gas had more than doubled compared to preindustrial era (to about 1,730 ppb); then it remained nearly constant for nearly a decade before a strong growth resumed in 2007 (Dlugokencky et al., 1998; Kai et al., 2011; Nisbet, Dlugokencky, and Bousquet, 2014). Assessing the role of natural gas in this increasing burden requires a stepwise approach: we must first establish the total of $CH_4$ emissions associated with the natural gas industry, compare it first to other anthropogenic source of methane and then to all biospheric fluxes of the gas, and, finally, quantify its contribution to the overall global warming potential (GWP). Only after going through this sequence we could determine the degree of concern attributable to methane emissions associated with gas industry. Obviously, it would be counterproductive if the industry producing the fossil fuel with the lowest specific $CO_2$ emissions would negate much, even most, of that desirable effect due to large volumes of methane that it let escape into the atmosphere.

If all natural gas would be recovered and burned, then its contribution to global warming would come only indirectly, due to the oxidation that transforms the simplest alkane into $CO_2$ and water. But it is inevitable that some methane escapes to the atmosphere before it can be burned, during the drilling of hydrocarbon wells, during their often decades-long operation, from the gathering pipelines in gas fields, and during the processing and long-distance transportation and local distribution to industrial users and households. Moreover, unwanted gas produced without any market access has been always deliberately flared, and this wasteful and environmentally damaging practice still continues at an unacceptable scale. And, finally, coal mining, thanks to China recently the most rapidly increasing kind of fossil fuel production, is also a source of uncontrolled releases of methane.

Because $CO_2$ remains by far the most important anthropogenic greenhouse gas, it has been a standard practice to express the warming impact of other gases in terms of $CO_2$ equivalent ($CO_2e$). This involves the concept of the GWP, a measure that expresses equivalent effects of various gases over a 100-year period. This standardization could be done for shorter (20 years) or longer (500 years) periods, and it is necessary due to different lifetimes of greenhouse gases in the atmosphere: $CO_2$ molecule may remain aloft for up to 200 years, methane's average is just 12 years, while nitrous oxide ($N_2O$) persists for more than a century. According to the IPCC's Second Assessment Report, the GWP for the simplest chlorofluorocarbon ($CCl_3F$) is 3,800 $CO_2e$; for $N_2O$, it is 310; and for methane, it is 21 (IPCC, 2007; 1 Mt $CH_4 = 21$ Mt $CO_2e$). For comparison, methane's potential for a 20-year period would be about 72, while that for 500-year span would be less than 8. I will quote published estimates of $CH_4$ emissions in Mt/year, and for comparisons with other greenhouse gases, I will use the standard mass conversion of (1 $CH_4 = 21$ $CO_2e$).

Natural gas industry generates $CH_4$ in several ways: as the gas escaping during the fuel's extraction; due to deliberate venting, while flaring (controlled combustion of escaping gas) produces $CO_2$ and is only a minor $CH_4$ source; and as losses during gas processing and pipeline transportation. $CH_4$ emissions from stationary combustion sources are highest for residential wood combustion (300 g/GJ) and (at just 1 g/GJ) negligible for either gas- or coal-based electricity generation (USEPA [United States Environmental Protection Agency], 2008). Our best understanding of production emissions comes from a large number of direct measurements at the US onshore sites (Allen et al., 2013). They included gas from well completion flowbacks (gas dissolved or entrained in liquids that must be removed from wells before the beginning of extraction), unloadings (lifting of accumulated liquids that restrict gas flow from producing wells), and routine operation of well sites (including wellheads, separators, controllers, and tanks).

Assuming that those measurements were representative of nationwide operations, the annual $CH_4$ emissions from these operations would be 957,000 (±200,000) t, while the estimate for comparable emissions categories in the national inventory was 1.2 Mt (USEPA, 2014). This is a fairly close agreement for inherently uncertain estimates of this kind. After adding USEPA's estimates for other source (not measured directly in their study), Allen et al. (2013) ended up with the total of 2.3 Mt or 0.42% of the nation's gross natural production, compared to the USEPA's 2011 total of about 2.55 Mt or 0.47% of the gross output. Processing losses are usually no higher than 0.2%, and leakage and venting during transmission and distribution range mostly between 0.3 and 1.0% of gross production.

Not surprisingly, natural gas losses from the world's longest pipelines connecting Western Siberia with Central Europe have received particular attention. According to Reshetnikov, Paramonova, and Shashkov (2000), during the early 1990s, their losses (established by inlet–outlet difference) amounted to 47–67 $Gm^3$ or 6–9% of total extraction. That would have been an extraordinary waste as the volume of 50–60 $Gm^3$ was, at that time, larger than Italy's total gas imports. In contrast, several Russian, American, and German measurements of the mid-1990s indicated losses of only about 1% of the gas produced, but they were questioned due to a small number of surveyed sites. The uncertainty was resolved by a comprehensive German–Russian measurement at compressor stations and along export pipelines in 2003. Their results indicated annual losses of 3.4 $Gm^3$ or an equivalent of 0.6% of produced gas or 0.7% when underground storage is included (Lechtenböhmer et al., 2007). The authors put the 95% confidence interval at 0.5–1.5% or about 7% of the total volume of energy used by Gazprom for pipeline operations.

The last source of emission from traditional oil and gas industry is methane flaring (see also Figure 2.2). Flaring that takes place in refineries and gas plants, as well as during well tests, is a minor source of emissions compared to the flaring associated (solution) gas produced along with crude oil or bitumen. Those emissions range between 1 and 4 kg $CH_4$/t of burned gas, and satellite observations are used to estimate annually flared volumes on national and global scales. The global level between 2009 and 2012 was around 140 $Gm^3$ or about 95 Mt/year (GGFRP, 2014). This means that flaring would be emitting at least 95,000 and as much as 380,000 t $CH_4$ a year, a negligible contribution that is far smaller than the error inherent in estimating emissions from all activities associated with natural gas production. The most realistic range of losses associated with properly operated natural gas production and transportation is thus between 1 and 2.5%, and with the 2010 global production of 3.2 $Tm^3$, that would prorate to roughly 32–80 $Gm^3$ or 22–54 Mt of $CH_4$.

Historic estimates of these emissions show that the industry remained a negligible source of $CH_4$ until after WW I, with annual rates increasing from less than 0.5 Mt $CH_4$ in 1900 to less than 8 Mt in 1950 and to about 33 Mt by 1990 (Stern and Kaufmann, 2001). Höglund-Isaksson (2012) put the global $CH_4$ emission from gas production and transportation in the year 2005 at about 29 Mt (with about 60% due to leaks during long-distance transmission and distribution), and calculations by the Global Methane Initiative (2014) have the 2010 emissions from the global oil and gas industry at roughly 65 Mt. But natural gas industry is just one-half a dozen anthropogenic sources of $CH_4$, and its emissions must be compared to those emanating from other activities, above all from agriculture.

Human intervention in the methane cycle began tens of thousands of years ago with deliberate burning of forests and grasslands; later (sometime between 8,200 and 13,500 years ago) came domestication of rice (Molina et al., 2011) that led to expanded anaerobic fermentation in water-covered soils and domestication of ruminants, starting with sheep in the Southwest Asia about 11,000 years ago, followed by goats and cattle at about 8,600 BCE (Harris, 1996; Chessa et al., 2009). Methane from coal mining began to add significant amounts of $CH_4$ during the eighteenth century, from urban landfills and municipal waste water during the nineteenth century.

Stern and Kaufmann (2001) prepared the most comprehensive inventory of historical anthropogenic emissions of $CH_4$ between 1860 and 1994. Their estimates for 1900 show the total of about 114 Mt, with nearly half from wet fields (mainly rice cultivation) and almost a third from livestock and roughly 10% each from the burning of phytomass and coal mining. By 1950, the total was up to nearly 180 Mt, with more than a third from wet fields, nearly 30% from livestock, 12% from coal mining, and still only less than 5% from natural gas production and flaring. The series ends in 1994 with the total of 371 Mt $CH_4$, with rice at 30%, livestock at 27%, and natural gas emissions at 9%.

Other studies of anthropogenic $CH_4$ emissions put the global total at the end of the twentieth century at between 300 and 350 Mt, and the USEPA's values (expressing the total in terms of $CO_2e$ with GWP = 21) were just over 300 Mt for the year 2000 and roughly 343 Mt for 2010 (USEPA, 2012). Höglund-Isaksson (2012) calculated 323.4 Mt in 2005, and the Global Methane Initiative (2014) put the 2010 emissions at about 327 Mt, with agriculture (mainly enteric fermentation of livestock and rice cultivation) contributing 50%, oil and gas industry 20% or about 65 Mt $CH_4$, and coal mining about 20 Mt or 6%. Recent estimates thus show a fairly good agreement given the multiple uncertainties inherent in quantifications of this kind and allow the following conclusions: in 2010, anthropogenic emissions of $CH_4$ were most likely no higher than 350 Mt, with natural gas responsible for no more than 70 Mt.

In turn, anthropogenic emissions of $CH_4$ are added to a few relatively large natural source that have been emitting the gas for hundreds of millions to billions of years. Beerling et al. (2009) published an interesting modeling exercise of tropospheric $CH_4$ levels during the past 400 million years. They scaled a wetland emission estimate for the middle Pliocene (3.6–2.6 million years ago) by the relative rate of coal basin deposition. This indicated extremely high peaks of about 12,000 parts per billion (ppb) during the Permo-Carboniferous era when tropical swamplands reached their largest extent and lows of just 100 ppb during the Triassic (coal-gap age). Prehuman biogeochemical cycle of $CH_4$ was dominated by emissions from natural wetlands (about 100 Mt $CH_4$/year), with termites and emissions from anoxic marine sediments each an order of magnitude lower.

The most comprehensive review of global methane sources and sinks during the recent decades (Kirschke et al., 2013) ended up with the following totals: natural emissions (mostly from wetlands) of 218–347 Mt $CH_4$/year (for, respectively, top-to-bottom and bottom-up accounting approach), anthropogenic emissions of 335–331 Mt, and 96 Mt from all fossil fuels (Figure 7.6). This means that human activities now account roughly for between 50 and 60% of all methane additions to the atmosphere, fossil fuels account for 14–18% of the total influx, and natural gas production and transportation for no more 10–13%.

But it should be noted that the range for global wetland emissions of methane (175–217 Mt/year) is as large as it was in 1974 when the world's first global methane budget was published, largely because annual natural methane fluxes can vary by a factor of two or more (Christensen, 2014). And a recent discovery of hundreds of ocean floor seeps leaking methane off the East Coast of the United States is yet another illustration of our inadequate understanding of natural $CH_4$ emissions. Skarke et al. (2014) identified about 570 gas plumes at depths of 50–1,700 m between Cape Hatteras and Georges Bank, a finding that suggests existence of as many as 30,000 of such vents worldwide—and their emissions have not been properly accounted for in any global $CH_4$ inventories.

Keeping that in mind, the final step in assessing the contributions of natural gas-related $CH_4$ emissions to global warming is to assign the share of radiative forcing attributable to the gas. In 2011, forcing's net value (after taking into account cooling effect of sulfates, nitrates, and organic carbon) was 2.83 W/m² (IPCC, 2013), with $CO_2$ contributing 1.82 (64%), $CH_4$ being a distant second with 0.48 W/m² (17%), and

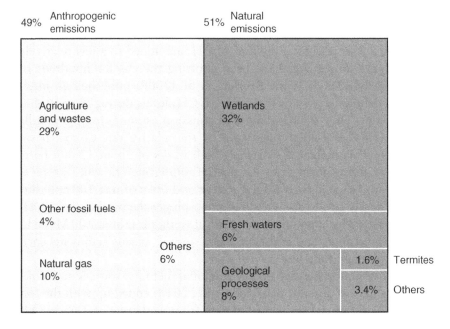

49% Anthropogenic emissions          51% Natural emissions

**Figure 7.6**    Global methane emissions.

$N_2O$ in the third place with $0.17\,W/m^2$ (6%). Assigning between 10 and 15% of methane's radiative forcing to emission from natural gas production and transportation makes the industry responsible for just between 0.05 and $0.07\,W/m^2$ or merely on the order of 2% of aggregate radiative forcing in 2011.

To complete the account of radiative forcing contributions attributable to natural gas, we must add $CO_2$ emissions from the fuel's combustion: they will, of course, make a much larger contribution than those attributable to methane. Until 1845, they accounted for less than 5% of the fossil-fuel total and hence were smaller than the inevitable calculation error (due mainly to changing carbon content of coal). By 1975, combustion of natural gas produced 13% of all $CO_2$ emissions from fossil-fuel combustion; by the year 2000, the share rose to 19% and by 2010 declined marginally, mainly due to rising carbon emissions from China's increasing coal combustion.

But because $CO_2$ has a relatively long atmospheric lifespan, we should consider the cumulative contribution as well. Between 1750 and 2010, combustion of fossil fuels, gas flaring, and cement production emitted the grand total of about 350 Gt C into the atmosphere (CDIAC, 2014) of which about 50 Gt C came from the burning of natural gas, the share roughly rounded to 15%, and the same share would be the most appropriate multiplier to estimate the carbon burden attributable to $CO_2$ emissions from natural gas. Multiplying carbon dioxide's rate of $1.82 \, W/m^2$ by 0.15 yields $0.27 \, W/m^2$, and combined with the methane's forcing share, it produces the combined contribution of about 0.3–$0.35 \, W/m^2$ or about 10–12% of the total anthropogenic radiative forcing, clearly an excellent performance for the fuel that now provides nearly 25% of the world's TPES!

At the same, it may be too optimistic to expect that increased use of natural gas will bring substantial decarbonization of the global energy use. Indeed, simulations by McJeon et al. (2014), based on five state-of-the-art integrated assessment models of energy–economy–climate systems and independently forced by an abundant natural gas scenario, show large increases in the combustion of gas (up to 170% more by 2050) with the resulting impact on overall $CO_2$ emissions ranging from a modest decline of 2% to an increase of 11% and with a majority of the models indicating a small gain (range of –0.3 to 7%) in climate forcing associated with the increased reliance on natural gas.

## 7.3.2 Methane from Shale Gas

New concerns about atmospheric impacts of natural gas (and oil) production have been created due to America's rapid expansion of hydraulic fracturing: its opponents realized that high methane losses during hydraulic fracturing, well completion, and transportation could reduce, or even negate, the benefits of gas as a fuel with much lower GWP than coal. I will offer a brief, and largely chronological, summaries of some recent claims and counterclaims about $CH_4$ emissions from shale gas extraction and transportation in the United States: these findings show how difficult it is to come up with any confident generalizations of fractional methane losses (be they from fracking operations or from conventional gas extraction) and how wide the range of uncertainty remains.

A study done at the National Energy Technology Laboratory used a detailed life-cycle model for natural gas that included about 30 items of processes encompassing extraction, processing, transport, and conversion, and it concluded that average natural gas (a mixture of conventional onshore, offshore, associated, tight, shale, and coal bed methane) base-load electricity generation has life-cycle greenhouse gas emissions 53% lower than average coal base-load power generation (Skone, 2011). Four other life-cycle analyses found similar, or even higher, benefit when comparing gas to coal (Burnham et al., 2011a; Hultman et al., 2011; Jiang et al., 2011; Stephenson, Valle, and Riera-Palou, 2011).

In contrast, Howarth, Santoro, and Ingraffea's (2011) study was the first that found the very opposite: they concluded (on the basis of limited measurements) that venting and leaks from shale gas production will produce, over the lifetime of a well, methane emissions at least 30% higher and perhaps more than twice as large as those from the production of conventional gas. They estimated life-cycle $CH_4$ emissions of a well to be 3.6–7.9% of produced gas for shale compared to 1.7–6% for conventional gas. Moreover, they also calculated that on the 20-year horizon, emissions from shale gas are also at least 20% higher than those from coal production and possibly more than twice as high as for coal when expressed per unit of energy available during combustion and that there is little or no advantage over coal even on the 100-year horizon. According to the study, offshore gas had the lowest GWP (but still about 25% higher than coal), and Barnett shale the highest (about 2.5 times as much as coal).

If true, then shale gas would lose all advantage in terms of carbon intensity of its production, although higher combustion efficiencies in household furnaces, large boilers, and gas turbines would still give it an overall, but relatively slight, edge. In response to a critique of their findings, Howarth, Santoro, and Ingraffea (2012) stood by their conclusions, and a study of air measurements at the Boulder Atmospheric Observatory tower, done to assess emissions from Denver–Julesburg Basin, ended up with a similar emissions range: Pétron et al. (2012) concluded that about 4% (range of 2.3–7.7%) of raw methane is leaking from infrastructure associated with gas production, a rate at least twice as high as has been generally assumed by the industry. These studies received wide attention because they suggested that methane leaks during gas production may entirely offset climate benefits of natural gas (Tollefson, 2012). Alvarez et al. (2012) concluded that emissions higher than about 3.2% of gas extraction result in immediate net radiative forcing worse than coal when the two fuels are used to generate electricity.

But the claims of gas being a worse choice than coal were questioned as responses to Pétron et al. (2012) pointed out that the conclusion depends on the way the measured concentrations were attributed and interpreted (Cathles, 2012; Levi, 2013). Most notably, different assumptions about air mixing with the stock tank and gas well leakages would produce $CH_4$ loss rates of just 1.5–2%. O'Sullivan and Paltsev (2012) reviewed emission data from about 4,000 horizontal fracked wells brought into production in 2010, and they highlighted the difference between potential and actual methane losses. While potential fugitive emissions averaged 228,000 kg $CH_4$ per well, the use of flaring and reduced emissions during well completion (capturing the flowback gas) lowered the actual emissions to about 50,000 kg $CH_4$ per well, much below some widely quoted estimates. This led them to conclude that hydraulic fracturing of shales has not substantially altered the overall intensity of greenhouse gas emissions from the US natural gas extraction.

Other studies confirmed the benefits of natural gas. Deutsche Bank Climate Change Advisors study of life-cycle greenhouse gas emissions used the USEPA's adjusted methodology, and it found that, on the average, US natural gas-fired electricity generation emitted 47% less greenhouse gases than coal (Deutsche Bank Group, 2011). And a comprehensive reevaluation of life-cycle greenhouse gas emissions of shale gas, conventional gas, coal, and petroleum—based on the USEPA's latest emission estimates and comparing the impact per MJ of fuel burned, per kWh of electricity produced, and per km driven for transportation services—concluded that that shale gas emissions are 6% lower than those from the conventional gas (but the difference is too small to make claim indisputable), 23% lower than from gasoline, and 33% lower from coal (Burnham et al., 2011a).

Another study published in the same year introduced three interesting comparisons of natural gas-switching scenarios (Alvarez et al., 2012). The study did not focus specifically on shale gas as it compared the benefits of switching from gasoline or diesel and from coal-fired to gas-fired electricity generation, but its conclusions provide fairly emissions threshold for the GWP benefit of such substitutions. If CNG were to reduce immediately climate impacts from heavy-duty vehicles, well-to-wheels $CH_4$ leakage must be kept below 1% of total extraction, and new gas-fired plants would be better than efficient, new coal plants only if the leakage in the entire natural gas system (from well to delivery) is kept to less than 3.2% of total production. The study also called for a much better understanding of methane leakage from natural gas infrastructure, an appeal strengthened by a review of two decades of publications on natural gas emissions in the United States and Canada (Brandt et al., 2014).

On one hand, this metastudy concluded that actually measured emissions are consistently higher than those indicated by standard emission inventories and factors (with measured/inventory ratios commonly around two) and that a small number of what they call "superemitters" could be responsible for very large shares of the leaked gas. On the other hand, the authors concluded that high leakage rates suggested in some recent studies are unlikely to be representative of the entire natural gas industry (if so, the emission associated with natural gas production would exceed the observed total excess $CH_4$ from all sources) and that assessments done for 100-year impact do not indicate any system-wide leakages large enough to negate climate benefits of substituting coal by natural gas.

This controversy will continue for years to come as even near-perfect measurements of emissions from a significant number of wells cannot be extrapolated with a high degree of confidence to quantify nationwide leakage or assumed to represent a similar group of wells to be completed in the near future. The latest inventory of US greenhouse gas emissions and sinks, published in a preliminary version in February 2014, puts the total 2012 emissions from the country's natural gas systems at 6.2 Mt $CH_4$ or 130 Mt of $CO_2e$, with transmission and storage releasing about a third of the total and field production about 30% (USEPA, 2014). Emissions of about 6 Mt $CH_4$ are equal to 8.5 $Gm^3$ or to almost exactly 1% of gross natural gas withdrawals of 836.3 $Gm^3$ in 2012. But just 2 months later, publication of a field study from the Marcellus shale region of Pennsylvania indicated some exceptionally high $CH_4$ fluxes from natural gas wells (Caulton et al., 2014).

The study relied on an instrumented aircraft to measure emission rates in southwestern Pennsylvania in June 2012. While a regional flux of 2.0–14 g $CH_4/s/km^2$ was in the same range as bottom-up inventory (2.3–4.6 g $CH_4/s/km^2$), emissions from seven well pads in the drilling phase averaged 34 g $CH_4/s$ per well, or two to three orders of magnitude higher than estimated by the USEPA for that stage of gas production. The studied wells accounted for only about 1% of all wells in the region, but their emissions produced 4–30% of the observed regional flux, confirming the previous conclusion about "superemmiters." These findings are in contrast with the latest US inventory of greenhouse gas emissions that shows $CH_4$ releases from natural gas systems declining by about 4% between 2005 and 2012 even as the gas extraction rose by 33% (USEPA, 2014).

Brandt et al. (2014) concluded that the US methane emissions are 25–75% (best estimate about 50%) higher than the USEPA's total but that higher emissions from hydraulic fracturing used in shale gas extraction account for only about 7% of the additional $CH_4$. Moreover, system perspectives may show that downstream leaks are no less important than the field emissions. Studies of methane emissions from pipeline and other leaks in Boston indicate higher than expected rates, and reducing those losses may be a key component of making shale gas less of a factor in any climate change concerns (Kintisch, 2014).

Recent quantifications of $CH_4$ releases from natural gas systems range from 0.6 to 11.7% of gas production for upstream (well sites) and midstream (processing) activities and from less than 0.1 to 10% for downstream (transmission, storage, distribution). But even if we assume that a high average of 5% were to apply to all natural gas production systems (and not just to hydraulic fracturing), then the 2013 global gas production of 3.37 $Tm^3$ would release nearly 170 $Gm^3$ (110 Mt) $CH_4$. That would be nearly 60% higher than the previously cited best estimate of $CH_4$ emissions attributable to natural gas—but the difference (40 Mt $CH_4$) remains considerably smaller than the lasting range of uncertainty regarding the natural emissions of the gas from wetlands (140 Mt) and there are numerous opportunities to reduce leakage.

More studies are unlikely to narrow the existing uncertainty of emission rates, and disputes about the life-cycle warming potential of gas systems in general and shale gas extraction in particular will continue. What is not at all in dispute are wasteful high emissions of $CO_2$ generated by gas flaring that have accompanied shale oil production in the United States. Development of the Bakken formation of North Dakota, the state whose crude oil output had increased more than 100-fold between 2003 and 2013 and that became the country's second highest oil producer after Texas in spring 2012 (USEIA, 2014l), is the most visible proof of this waste. Extraction of Bakken shale oil is associated with large volumes of liquid-rich natural gas, and construction of new gathering line and trunk pipelines has not kept pace with the rapid rate of drilling new wells and producing new crude.

As a result, nighttime satellite images show that gas flaring in northwestern North Dakota has created a patch of light whose intensity is not quite as bright as those of the nearest large cities, Minneapolis in Minnesota and Denver in Colorado—but whose size is considerably larger (Figure 7.7). But this wasteful and environmentally harmful reality is no generic and lasting indictment of hydraulic fracturing, merely a result of specific circumstances resulting in excessive temporary flaring that will be reduced (North Dakota's new standards approved in July 2014 will require capture of 90% of all gas released during drilling) and eventually eliminated as new pipelines enter into service.

Comparisons of natural gas losses associated with conventional drilling and hydraulic fracturing show no or minimal difference for most routine activities, provided that all of them are conducted with care, a condition that is not always present in early stages of resource extraction

Figure 7.7    Flaring in Bakken. © NASA.

of which the rapid development of Bakken shale has been a notable
example. And compared to often long-lived conventional wells, the
exponential decline in well productivity in fractured shale formations
will require more frequent drilling to maintain a given level of output
and hence present more opportunities for methane leaks. But these challenges have technical solutions, and there are no insurmountable problems that would prevent methane emission from shale gas production to
be comparable, or lower, than those from conventional operations.

   And before leaving the topic of atmospheric emissions associated with
natural gas production, I must also mention that in some area there may
be also seasonal impacts associated with the releases of nitrogen oxides
(NO and $NO_2$) and volatile organic compounds. These air pollutants are
the essential precursors of photochemical ozone formation, and Edwards
et al. (2014) found that in Utah's Uinta Basin this leads to winter ozone
mixing ratios well in excess of present air quality standards.


## 7.3.3    Water Use and Contamination

Making clear-cut conclusions concerning the consumptive use of water,
contamination of drinking water aquifers and injection of salt-laden
water into deep wells is even more difficult than judging the
atmospheric impacts. HVHF is much more demanding and has a
much greater impact than the extraction of conventional natural gas

resources, but water requirements by HVHF need careful qualifications. A simple estimate shows that water needed for HVHF adds up to a tiny share of total US water withdrawals. In the United States, usual ranges of overall water requirements are given either as 2–8 million gallons or, more narrowly, as 4–6 million gallons per well, that is, as little as 7.5 and as much as 33 Ml (or 7,500–33,000 m³), compared to 1 million gallons (3.8 Ml) for a vertical well. Obviously, fracking longer lateral wells increases water demand, generally by about 18,000 l for every additional meter (Fractracker Alliance, 2014).

Average of 15 Ml (15,000 m³) would be perhaps the most representative requirement if a single value is needed to make proper order-of-magnitude estimates of aggregate water use. For about 30,000 wells (maximum drilled recently), that would amount to about 450 Mm³ or less 0.1% of all US water use in 2005 (Kenny et al., 2009). The shares are similarly low even in regions of concentrated HVHF: in the Marcellus formation in Pennsylvania, water consumption for fracking is only about 0.2% of total annual water withdrawals (Vidic et al., 2013). The share may be lower in rainy watersheds, but it could be much higher in arid regions. In Texas, HVHF uses less than 1% of the state's water withdrawals but with a large share of the state being so arid that may still tax local water supplies.

Yet another perspective is to consider water intensity of the most common alternative. Because coal will be the fossil fuel most commonly displaced by natural gas, it is instructive to compare water intensity of coal-based electricity generation with water intensity of shale gas-based generation. Scanlon, Duncan, and Reedy (2013) did precisely that for Texas (the state where concerns about water supply are particularly acute), and their conclusion is impressive: water saved by using gas-fueled CCGT instead of coal steam turbine plants is 25–50 times the volume of water used in hydraulic fracturing to extract the gas.

What makes the water use in fracking so demanding is that a requisite volume must be delivered to a single site within a brief period when fracking operations take place, that in many arid environments even a much smaller volume could not be secured from nearby sources, and that the delivery is nearly always an environmentally stressful experience. There are, of course, many single-well operations, but multiwell pads are becoming more common, and hence, the actual volume required at a single site may be commonly three to ten times as large (extreme range being 2-20 wells). Assuming the average requirement of 15,000 m³/well (mostly taken from streams and ponds, less frequently from municipal supplies and from recycled flowback liquid), a five-well pad site would need 75,000 m$^3$. With 25 m$^3$ of volume per truck, that would require 3,000 truck trips to a drill site, with noisy heavy diesel-powered trucks destroying rural roads (Figure 7.8).

Even more importantly, many water uses in modern society are non-consumptive, and that includes the largest industrial use for the cooling of condensed water in thermal electricity-generating plants: except for a small share lost to evaporation, water is returned, slightly heated, to lakes, ponds, or streams. And while other uses are consumptive—evapotranspiration and leaching removes irrigation water from crop fields—that water remains part of the rapid global water cycle as it gets precipitated (close or far downwind) or reenters aquifers. In contrast, water used for fracking is not only consumed in the process but ceases to be a part of rapid water cycle. Most of the loss is immediate as usually at least 70% and commonly up to 90% of water used for fracking remains deep underground within shale formations. In many cases, the fate of this water is uncertain: it may be absorbed by shales, but in some settings, it may be also a potential contaminant of aquifers.

Once HVHF is completed and pressure is lowered, a portion of fracking liquid mixed with formation water—anywhere between 10 and 50% of the volume initially pumped into the well, with modal flows close to 10%—returns to the surface (most of this flowback takes place in the

Figure 7.8   Heavy truck carrying fracking liquid. © Corbis.

first 2–3 weeks) and has to be hauled away for recycling (necessitating hundreds of additional truck trips from a drill site). But, unlike water polluted with organic substances, flowback fracking liquid cannot be recycled infinitely (its concentration of salts becomes eventually too high, and sometimes, even its initial contamination is too severe to be handled by standard water treatment plants), and it has to be disposed of, usually by injections into deep wells (far below drinking water aquifers) and hence again taken out of water cycle. In addition, drilling of a single horizontal well can yield on the order of 100 t of organic-rich cuttings whose dumping would acidify affected waters and leach heavy metals; moreover, the cuttings may also have impermissibly high levels of radiation and must be disposed in secured landfills (USGS [United States Geological Survey], 2013).

Water dominates the volume of fracking liquid (typically 90%). If varieties of sand, the most common proppant (an ingredient introduced to keep fractures open and hence to enable gas and oil flow, and usually amounting to about 9% of fracking fluid mass) were the only other ingredient, then concerns about fracking and management of waste water would be much less challenging. But all fracking liquids contain numerous additives: about 750 substances have been used by the industry, and many fluids contain scores of different chemicals. Composition of fluids used in different regions is available from Halliburton (2014). Although their relative volume is small (typically <0.5%), their total volume used per fracking operation may be up to 150,000 l (150 m³).

Common additives include acids (helping to dissolve minerals and initiate tiny cracks), corrosion and scale inhibitors (methanol, formic acid), friction reducers (mostly petroleum distillates), gelling agents (to thicken fluid and help to suspend proppants, commonly guar gum and petroleum distillates), surfactants (to reduce the fluid surface tension, usually ethanol or methanol), pH adjusters, and biocides (FracFocus, 2014). Biocides (glutaraldehyde, ammonium chloride) are needed because stream or pond water contains bacteria whose growth could be enhanced by organics added to fracking fluid and result in emulsion problems, plugging, and souring (Biocides Panel, 2013).

Claims of contaminated drinking waters have come from a number of states (Arkansas, Colorado, New York, Pennsylvania, Texas, Virginia, Wyoming) and have been a major rallying point for the opponents of fracking. Reading their pamphlets and web pages, one would think that all methane is coming out of all faucets in the regions with HVHF. At the same time, it appears that some homeowners living less than 1 km from a shale gas well have a higher probability of having their drinking water contaminated with stray gases: that was the conclusion made by Jackson et al. (2013) after they analyzed 141 drinking water wells in northeastern Pennsylvania by correlating natural gas concentrations and isotopic signatures with proximity to shale gas wells. They detected methane in 82% of drinking water samples, and its average concentrations were six times higher in homes less than 1 km from natural gas wells, but, in a majority of cases, below the level that would require action for hazard mitigation.

In contrast, nine months of monitoring tracers injected into six shale gas wells in Pennsylvania's Marcellus shale showed no signs that fracking fluids are ascending toward the surface where they could contaminate well water (Stokstad, 2014). Contradictory findings are not surprising because the impact of shale gas extraction on regional water quality is not a topic for easy generalizations and simplistic conclusions (Vidic et al., 2013). Low, naturally occurring volumes of $CH_4$ (<10 mg/l) pose no health hazards, higher levels can increase the solubility of arsenic and iron and stimulate the growth of anaerobic bacteria, and high accumulations could result in an explosion.

Stray methane can enter wells from natural biogenic or thermogenic sources as well as from such anthropogenic sources as landfills, coal mines, pipelines, and abandoned old oil and gas wells (of which there are hundreds of thousands in Pennsylvania or Texas)—but, undoubtedly, the gas can also come from fractured casing and leaking cement seals that fail to insulate wellbores from aquifers that normally lie at least many hundreds of meters above the formations subjected to HVHF (Vengosh et al., 2013). Typical incidence rate of casing and cement problems is 1–2% but Vidic et al. (2013) found the frequency of 3.4% in Pennsylvania's Marcellus formation. Pinpointing methane sources in a well may be difficult, and moreover, in some regions (notably in Marcellus shale in Pennsylvania), intensive fracture networks may provide hydraulic connectivity between deep shale gas formations and the overlying shallow aquifers.

There are reasons for encouragement but also arguments for caution. After more than one million of HVHF treatments, there is perhaps only one documented case of direct groundwater pollution resulting from injection of fracking chemicals—but confidentiality requirements (imposed by ongoing investigations), expanding scope of fracking, and possibility of cumulative, slowly developing impacts are arguments for caution (Vidic et al., 2013). There is no doubt that in all regions with extensive HVHF the extraction will have widespread acceptance only with strictly enforced regulations designed to prevent aquifer contamination.

Some of these concerns may have been exaggerated, public opinion has been manipulated by dubious claims, and many early problems should be reduced or eliminated as operating experience mounts and as innovative solutions, including the use of nontoxic fracking mixtures and the least offensive biocides, get widely adopted. In 2013, Halliburton introduced CleanStim, a mixture (not surprisingly, more costly than typical products) that uses only additives approved by food industry, and a year earlier EPA- and FDA-approved SteriFrac, a pH neutral biocide to kill bacteria without any toxic by-products. Surface spills of fracking liquid can be reduced by secondary containment including dikes and berms, retaining walls with heavy plastic liners (Powell, 2013). And waterless fracking, using LPG, may be the ultimate technical solution (Loree, Byrd, and Lestz, 2014).

We already have enough evidence to state with confidence that hydraulic fracturing will not invariably poison the air, will not cause everywhere spates of localized earthquakes, and will not produce flaming faucets in all nearby areas (to list just the key frightening impacts attributed to it by its opponents)—but there is also no doubt that hydraulic fracturing, done in thousands of hurried repetitions and sometimes without adequate planning, has the potential to be often not just unpleasant and disruptive but even outright damaging at the local level. Many people simply want to know more about the true risks of hydraulic fracturing: in September 2013, the Pew Research Center found 49% of Americans opposed to the increased use of the activity, while, a year after the Fukushima nuclear disaster, 58% of Americans also opposed the increased use of nuclear power (Pew Research Center, 2013). Such perceptions cannot be simply dismissed, and energy companies must address them and explain the true risks involved: recently Exxon, now the United States' largest natural gas producer, promised to disclose more of such information.

In its media advertisement during 2014, Exxon stressed that "an amazing resource for Americans" can "be produced safely, while protecting water supplies" because there are "thousands of feet of protective rock between the natural gas deposit and any groundwater," and "in addition, multiple layers of steel and cement are installed in shale gas wells to keep the natural gas and fluids used in the production safely within the well." All true, but as I have just explained, neither reality makes contamination impossible, and we will need a longer record of cumulative performance (at least another 5 years) before we will be able to appraise with confidence the real degree of risk posed by HVHF to water supplies.

The best way to resolve any disputes concerning the impacts of fracking is to establish baseline conditions of water and air quality prior to drilling and fracturing, a step recommended by the International Risk Governance Council (IRGC, 2013). While it may be too late to do that as far as many intensively developed areas in the United States are concerned, future developments would clearly benefit from such measures. And careful assessment will be imperative in many of the world's regions whose hydrocarbon-bearing shales are in arid regions where any fracking could lead to high or even extremely high stress on local water resources. Global assessment by Reig, Luo, and Proctor (2014) shows China and India in high-impact and Pakistan and Mongolia in extremely high-impact category.

Before I close this environmental section, I should note two important environmental advantages of gas compared to coal and oil. Unlike in the case of crude oil and refined oil products, where trucking and railroad accident and pipeline ruptures can lead to usually limited but still harmful spill, transportation of natural gas has no effect on water quality as any leaked fuel escapes into the atmosphere. And although the land claims of natural gas extraction are generally comparable to those of oil production, they are much smaller than for modern coal extraction, and in many countries producing all of these energies, they are the smallest of the three fossil fuels. These spatial claims are best quantified by calculating prevailing power densities, expressed as annual flux of energy per unit of surface ($W/m^2$).

As already noted (in Chapter 3), power densities of natural gas extraction are typically between $10^3$ and $10^4 W/m^2$ and (commonly 2,000–12,000 $W/m^2$), and this means that a gas field producing annually 1.2 $Gm^3$ of natural gas (an equivalent of 1 Mt of crude oil) will have a land footprint as small as 10 ha or as large as 70 ha: even the latter total is a square with sides of less than 840 m. Gas output from hydraulically fractured horizontal wells has rapid production decline with power densities falling from $10^3 W/m^2$ in the first year to low $10^2 W/m^2$ just a few years later. Processing of natural gas has high-throughput densities of $10^4 W/m^2$, and power densities of long-distance pipeline transportation range from $10^2$ to $10^3 W/m^2$.

As expected, power densities of oil extraction are similar, with the North American oil fields having long-term cumulative rates of about 2,500 $W/m^2$ for more than 80 years in California and about 1,100 $W/m^2$ for 50 years in Alberta, with the peaks of $10^4 W/m^2$ for the world's most productive Middle Eastern fields (Smil, 2015). Only the largest underground coal mines tapping thick seams have power densities in excess of 10,000 $W/m^2$;

smaller operations are often above $2000 \text{W/m}^2$ and usually not below $1000 \text{W/m}^2$. The largest surface mines that extract thick seams of bituminous coal have power densities of more than $10,000 \text{W/m}^2$, but most common operations rate $1,000–5,000 \text{W/m}^2$, and for Appalachian mountain-top removal, it can be as low as 200 (or even below 100) $\text{W/m}^2$.

Finally, I should note an indirect environmental impact of natural gas extraction that could have been expected but whose intensity in Oklahoma still came as a surprise. Localized earthquakes have been attributed to deep well injections of water, mainly from dewatering operations used to separate oil and gas from large volumes of target rock formation and to a lesser extent from recent injections of contaminated fracking water across Ohio, Arkansas, Texas, and Oklahoma (Hand, 2014). Van der Elst et al. (2013) found that areas with suspected anthropogenic earthquakes are also more susceptible to earthquake triggering from natural transient stresses generated by the seismic waves of large remote earthquakes and that fluid injections can bring critically loaded faults to a critical state.

As a result in 2013 and 2014, Oklahoma registered a similar, or higher, number of magnitude 3 or greater quakes than California. Damage to structures has been in most cases minimal, but the largest event (magnitude 5.7 in 2011 in Prague, OK) damaged four spires on Benedictine Hall at St. Gregory's University in Shawnee, OK, located 25 km from the epicenter. Given that the likelihood of larger quakes is correlated with the frequency of smaller quakes (for every 100 magnitude 5 events, there will be 10 magnitude 6 quakes), there is a legitimate concern about triggering a much stronger anthropogenic earthquake.

# 8

# The Best Fuel for the Twenty-First Century?

Those who have written effusively about America's recent natural gas-based energy revolution, or renaissance, boom, or bonanza that will change the world—and Gold (2014) packed it nearly all into the title and subtitle of his book: *The Boom: How Fracking Ignited the American Revolution and Changed the World*—would say that the title of this closing chapter should not have a question mark (implying at least some uncertainty, if not a real doubt) but that it should be a simple, affirmative statement. And two prominent American promoters of natural gas called for the fuel's new central role even before the expansion of shale gas production showed the potential of this new source of energy and led to an impressive decline of gas prices: in 2008, they unveiled similarly bold plans that envisaged a large-scale reorientation of American economy from oil to gas.

The plan that got much more publicity (because his author was willing to spend a great deal of his own money on advertising it and was always available to promote it personally in media) was revealed during the summer of 2008 by T. Boone Pickens who made his considerable fortune first in Texan oilfields and then as a corporate raider (Pickens, 2014). The 10-year plan rested on twofold substitution: to begin with, Pickens called for making the Great Plains "the Saudi Arabia of wind power" and to use this new wind power to replace all electricity produced by natural gas combustion and, in turn, to compress the available natural gas to be used in clean, efficient vehicles.

*Natural Gas: Fuel for the 21st Century*, First Edition. Vaclav Smil.
© 2015 John Wiley & Sons, Ltd. Published 2015 by John Wiley & Sons, Ltd.

That switch was to reduce America's dependence on imported crude oil by more than a third (thus strengthening the country's fiscal position) and to create a new, large-scale domestic industry that would bring much needed jobs and economic revitalization to the Great Plains, a region of ever larger farms and ever smaller population. Pickens presented his plan to the Congress and spent $58 million on advertisement to generate public support. His principal motivation was to curtail what he saw as America's addiction to oil that poses threat to "our economy, our environment and our national security" and that "ties our hands as a nation and a people" (Pickens, 2008).

Before the second switch (gas for liquid automotive fuels) could be put in place, his plan called for building more than 100,000 large-capacity wind turbines and at least 65,000 km of high-voltage transmission lines to connect new generating capacities in Texas, Oklahoma, Kansas, Nebraska, and Dakotas to large coastal cities. Pickens estimated that his plan will need roughly $1 trillion in private investment for new wind power and at least another $200 billion for the requisite transmission lines. And that would have to be followed by conversions of tens of millions of cars to natural gas fuel. Just months after launching the plan, Pickens (before the end of 2008) acknowledged that the need for new transmission lines makes any rapid progress impossible, and he also changed his gas-switching proposal from cars to trucks (that would require fewer filling stations selling compressed natural gas).

Combination of economic downturn (it began just a few months after the plan was launched) and rapidly rising availability of inexpensive natural gas from hydraulic fracking ended the short-lived plan. In July 2009, Pickens still claimed that "Financing is tough right now and so it's going to be delayed a year or two" (Rascoe and O'Grady, 2009, 1), but soon afterward, even his own key wind power project (a 4 GW wind farm, the world's largest, near Pampa in Texas) was first delayed and then, in January 2010, canceled because the $4.9 billion worth of the needed transmission lines would not pass all regulatory requirements before 2013. Between 2010 and 2013, as shale gas extraction expanded and as natural gas prices fell, little was heard about Pickens and his plan.

But Pickens was back in July 2014 when he told CNBC (Belvedere, 2014, 1) that

we're down to 4 million barrels a day of OPEC oil (from 7 million) ... and we can knock that out within the next three years.... All you have to do is switch natural gas over to the heavy-duty trucks.... If [Washington] had gone with me six years ago, you figure you could probably had the job done in three to four years.... If that were the case, you would have cut out 75 percent of OPEC, because 8 million trucks converted to natural gas off of diesel is 3 million barrels a day.

But in July 2014, there was still no large-scale program to switch heavy trucks to natural gas, and the chances of Pickens plan were no better than in 2008.

The second transformative plan was offered by Robert A. Hefner III, an Oklahoma oil and gas developer and the founder and owner of The GHK Company who pioneered ultradeep natural gas exploration and completed the world's deepest and highest-pressure natural gas wells. Hefner's work as gas explorer and producer led him to believe—already during the 1970s, decades before a new consensus began to emerge— that the US reserves of the fuel may be larger than even the country's huge coal deposits, and he thought that the best use of this rich resource is to convert a large share of the US vehicle fleet to natural gas. Already in 1989, Hefner argued that "the Detroit of natural gas fueled vehicles in the future should be located in Oklahoma … Natural gas will be the principal energy to fuel the U.S. and global economies into the twenty-first century," and in 1991, he told an interviewer that "the nation has enough natural gas to last 100 years even without imports..... We can convert a third of our cars and trucks to natural gas" (GET, 2014).

Hefner's Grand Energy Transition (GET), a comprehensive version of his plans, was finally published in 2009 (Hefner, 2009); its subtitle, *The Rise of Energy Gases, Sustainable Life and Growth, and the Next Great Economic Expansion*, made it clear that the author saw the rise of natural gas as the key unlocking unprecedented environmental, economic, and quality of life benefits—but, in essence, his plan was fundamentally the second part of the Pickens scheme, although with some notable tax adjustments. Hefner argued that the conversion would be relatively easy because the requisite infrastructure is already in place: most of the existing gasoline filling stations are already connected to gas distribution lines as are most of America's homes whose owners could fill more than 130 million vehicles after installing convenient home-fueling devices.

Hefner's calculations had crude oil imports falling by about 250 Mt/ year with cumulative saving of trillions of dollars in a decade, with concurrent benefits of some $100 billion of private investment, and about 100,000 new jobs spurred by the rising demand for natural gas. But unlike the Pickens plan, Hefner's GET was to rely on a fundamental tax restructuring that would eliminate taxes on labor and capital and replace them with levies on coal and oil (a green consumption tax). Success of Hefner's plan was thus dependent not only on infrastructural changes and technical adjustments but also on the readiness of the US Congress to go along with his bold new tax schemes. The latter move had little chance to be enacted by now chronically dysfunctional Congress, and the Hefner plan has not been any more successful than the Pickens plan. This is not at all surprising as energy transitions are generally gradual, often protracted, processes (Smil, 2010a).

Their incremental progress can be accelerated or retarded by specific policies—but only rarely do such measures result in truly revolutionary shifts: energy systems are too complex and generally fairly long-lived and hence too inertial to be rapidly redirected by deliberate action designed to change their fundamentals. Grand plans aimed at their basic redesign thus have a very low probability of success, and we are left trying to do the best we can to nudge the process in what we think is the best direction—but we still must keep in mind that, in retrospect, we may find such actions not as beneficial as we thought them to be at the beginning. In this book, I have endeavored to make a strong case for natural gas as a preferred fuel—but as a lifelong student of energy in modern society I also know that there are no ideal fuels, and as an inter-disciplinary scientist trying to understand behavior of complex system, I always try to think about unanticipated impacts, unintended consequences, and counterintuitive outcomes.

That is why I insist on the question mark in the chapter's title—and on another question mark in trying to outline how far the extraction and conversion of natural gas might go during the coming decades, more specifically during the first half of the twenty-first century. My intent is not to answer the question either by a comprehensive review of the recent US and global forecasts of natural gas extraction or by offering my own forecasting model or at least some probable ranges of future volumes. I have shown that energy forecasting, in common with nearly all other long-range forecasting endeavors, has a poor record (Smil, 2003, 2010a), and hence, it is best to avoid additional contributions to this futile intellectual effort. Instead, I will concentrate on exploring the range of the most likely outcomes based on the best physical and technical evidence.

## 8.1   HOW FAR WILL GAS GO?

Despite its dismal record, long-range forecasting—as attested by an endless procession of new studies and models—retains an irresistible appeal. I will review some of these recent efforts in order to illustrate their questionable nature, but first, I will address the validity of an influential forecasting technique that appeared to solve the intractable challenge of impartial forecasting by relying on a fairly simply numerical approach that appeared to yield, repeatedly and for a wide variety of phenomena, results that were closely corresponding to past realities and hence seemed to be by far the best way to predict future outcomes.

In 1974, Cesare Marchetti, an Italian physicist and nuclear scientist, began to work at the International Institute of Applied Systems Analysis (IIASA), and during the late 1970s, as energy supply became a great public and scientific concern, he looked for a quantitative model that would describe energy substitutions, in both global and national terms, and he analyzed the market shares of successive uses of fuels and primary forms of electricity by using the Fisher–Pry model that was developed to study the market penetration of new techniques (Fisher and Pry, 1971). The model's fundamental assumptions are that technical advances can be seen as competitive substitutions and that the rate of fractional substitution is proportional to the remainder that is yet to be substituted. Because the adoption rates tend to follow a logistic curve, it is easy to calculate the market fraction ($f$) of a new technique (or a new energy source) and express it as $f/1-f$—and when that quotient is plotted on a semilogarithmic graph, it appears as a straight line.

The method was originally applied to many simple two-variable substitutions (including synthetic vs. natural fibers, plastics vs. leather, electric arc furnaces vs. open hearth steelmaking, etc.), but Marchetti used it for historical energy substitutions that begins with just two sources (as coal is taking market share from traditional biofuels) but now, at the global level, includes six primary sources (biofuels, coal, oil, natural gas, hydroelectricity, and nuclear electricity). The first publication of these substitutions included a much-reprinted graph of the world's rising and falling primary energy sources between 1850 and 2100, and Marchetti (1977, 348), without any reservations, claimed that

the whole destiny of an energy source seems to be completely predetermined in the first childhood ... these trends ... go unscathed through wars, wild oscillations in energy prices and depression. Final total availability of the primary reserves also seems to have no effect on the rate of substitution.

In a subsequent, and more detailed, publication, he stressed that, despite such major perturbations as wars and economic crises and booms during the first three-quarters of the twentieth century, the penetration rates remained remarkably constant, as if "the system had a schedule, a will, and a clock" and hence is able to reabsorb all perturbations "elastically without influencing the trend" (Marchetti and Nakićenović, 1979, 15). This unbounded determinism seemed to offer an unexpectedly reliable means of long-range forecasting of global and national energy developments. But there were two problems with this simplistic trust: Marchetti's analysis contained several obvious errors, and almost as soon as it was published, new realities caused major deviations from expected substitution patterns.

The original analysis left out hydroelectricity, the most important source of primary electricity ever since the 1880s, and it used inexplicably low estimates of historical global fuelwood consumption that pointed to wood accounting for less than 1% of the total primary energy supply before 1995, while traditional biofuels still supply at least 10% of the world's primary energy use. And the simplistic belief in a built-in clock entirely missed the post-1970 deviations from supposedly set trends of falling coal shares, peaking and falling oil shares, and strongly rising gas shares. Marchetti's clocklike substitution mechanism was to deliver these global shares by 2010s—less than 5% for coal, about 25% for oil, and just over 60% for natural gas—while the actual shares were, respectively, 29, 34, and 24% (Figure 8.1). Some clock!

And, an important point when trying to discern the future of natural gas, the model's greatest error was in highly overestimating its future

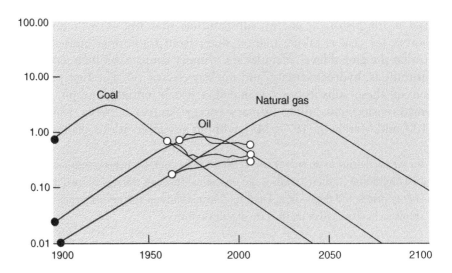

**Figure 8.1** Marchetti's fuel transitions and reality.

rise. At roughly 60% of global primary energy use, the gas would have delivered about 300 EJ of primary energy in 2010—while in reality, it provided no more than 120 EJ. Clearly, the system does not behave in a predetermined fashion, and natural gas consumption has shown the greatest deviation from the expected trend. The only correct feature of Marchetti's model is that it captures the gradual nature of energy transitions, with primary resource substitutions taking many decades before claiming significant shares of the overall market.

As already explained (in Chapter 7), the worldwide transition to natural gas has proceeded slower than the two preceding shifts. Natural gas reached 5% of the global primary energy supply by about 1930, rose to 10% by 1950, and to 20% only by 1995, and, using British Petroleum (BP) data series, it has yet to reach 25% (it was short of 24% in 2013). Consequently, it will have taken natural gas more than 80 years to go from 5 to 25%, while coal took 35 years and crude oil 40 years to span those shares. Of course, this slower rate of penetration is a function of scale: as the modern global primary energy supply expanded (in 2013, it was, excluding traditional biofuels, about 535 EJ; in 1950, it was about 80 EJ; in 1900, it was only 22 EJ), it is more difficult to claim additional share of the market. Adding 1% in 1950 required 730 PJ (equivalent of about 17 Mt of oil), and in 2013, it demanded 5.3 EJ or 127 Mt of oil equivalent. Clearly, high shares dictated by a dubious clock are not going to materialize: there is absolutely no chance that natural gas could supply 60% of the world's primary energy even by 2050 and a very low probability that it could make it as high as a third of the total by 2040.

While I will not offer any new forecasts, I think I should note at least five recent projections, three done by the world's largest oil and gas companies and the other two by leading energy bureaucracies. ExxonMobil (2013) looks as far as 2040 when it sees global energy supply at 743 EJ, which is about 35% above the 2010 level, with natural gas supplying 27% compared to 22% in 2010, or in absolute terms going from about 121 to 199 EJ, a 65% gain. The study also estimates sectoral consumption shares and sees no change for natural gas for residential and commercial use (supplying 22% in both 2010 and 2040), slight gains in industry (from 23 to 27%), and electricity generation (from 25 to 29%)—and still virtually no role in transportation: ExxonMobil does not foresee any large-scale CNG- or LNG-fueled road and maritime traffic or any massive GTL industry to supply the transportation sector.

BP, Exxon's smaller rival, chose 2035 as the last year of its forecast, and it has rightly stressed that fuel shares evolve slowly, and as oil continues to decline and gas continues to rise:

> by 2035 all the fossil fuel shares are clustering around 27%, and for the first time since the Industrial Revolution there is no single dominant fuel. Taken together, fossil fuels lose share but they are still the dominant form of energy in 2035 with a share of 81%, compared to 86% in 2012. (BP (British Petroleum), 2014a, 17)

The BP sees transport as the sector with the fastest growth of natural gas consumption (7.3%/year) but, of course, from a small base; in absolute terms, it sees most of the demand growth coming from industry and electricity generation (both rising by nearly 2%/year).

The BP expects shale gas to provide nearly half of the extraction growth in global gas, with the US shale gas (now more than 99% of the global total) still accounting for 70% in 2025, followed by 13% from China. Another forecast in accord with general expectations is that natural gas trade expansion will be dominated by Asia-Pacific region whose imports should overtake Europe by 2026 and could more than triple by 2035 to account for half of all net interregional imports, while the growth of shale gas changes North America from a net importer to a net exporter of gas as early as 2017. Imported gas should supply 34% of all consumption in 2035, slightly above the 2012 share of 31%; pipelines will remain the dominant means of delivery even as the share of LNG trade rises significantly, from 32% in 2012 to 46% of all exported gas by 2035.

And Royal Dutch Shell (2013) goes as far 2060 in its latest duo of scenarios called Mountains and Oceans. These two scenarios represent continuation of status quo, moderate economic growth and vigorous development of new gas resources on the one hand and a more constrained expansion of gas on the other. In the first case, global primary energy use rises to 992 EJ by 2060, with gas delivering 24%, slightly behind coal (25%) but nearly twice as much as oil (13%); in the second case, the global total is slightly higher at 1056 EJ, but gas supplies only 17% of the total, with oil and coal roughly even at 19%. For comparison, Shell's 2040 primary energy total is 822–856 EJ, at least 10% higher than 743 EJ estimated by ExxonMobil.

As for the two leading energy bureaucracies, the International Energy Agency issued a lengthy natural gas report in 2012 concluding the fuel "is poised to enter a golden age, but will do so only if a significant proportion of the world's vast resources of unconventional gas ... can be developed profitably and in an environmentally acceptable manner"

(IEA, 2012, 9). The promise can be achieved only if the social and environmental concerns associated with gas extraction are fully addressed by advancing production techniques that improve performance and assure public confidence in the new extraction processes.

These qualifications echoed the conclusion published a year earlier by the US National Petroleum Council:

> realizing the benefits of natural gas and oil depends on environmentally responsible development. In order to realize the benefits of these larger natural gas and oil resources, safe, responsible, and environmentally acceptable production and delivery must be ensured in all circumstances. (NPC, 2011, 8)

The IEA called specifically for stronger regulation of fracking, including improved disclosure of chemical composition of fracking fluids, better monitoring of water discharges and wastewater disposal, and further research on environmental impacts.

If these golden rules regarding "full transparency, measuring and monitoring of environmental impacts and engagement with local communities are critical to addressing public concerns" are followed, the IEA estimates that they would raise the cost of a typical shale gas well by 7%, but higher fuel availability would have a strongly moderating impact on gas prices, and the global demand for natural gas would rise by more than 50% between 2010 and 2035, and by that year, gas would supply 25% of the worldwide primary energy demand to become the second most important source after crude oil.

Finally, in its reference forecast, the USEIA puts the share of natural gas at 24% of the global primary energy use in 2030 (only marginally above the 2010 share) and at 22% by 2040, while for North America, it forecast the rise from 25% in 2010 to 28% by 2030 and 29% by 2040 (USEIA (US Energy Information Administration), 2014i). During the shared forecasting time span (up to 2030 or 2040), all of these five forecasts suggest very similar, and generally linear, growth of global natural gas consumption that would roughly double the 2010 volume by 2030 and raise the fuel's share in total primary energy consumption to just short of 30%. Highly conservative nature of all of these forecasts may be a surprise only to those uncritical observers who believe (contrary to rich historical evidence) that fuel totals and fuel shares can shift rapidly at the global level.

And perhaps the most comprehensive study of US natural gas prospects concluded that, based on reserves of $60\,Tm^3$, most of the increased extraction will go for electricity generation, and that if energy prices were to reflect specific carbon burdens natural gas would almost completely displace combustion of US coal by the year 2035, but its production would start declining by 2045 as the global energy system moves toward no-carbon alternatives (MIT (Massachusetts Institute of Technology), 2011). More specifically:

> under the mean resource estimate, U.S. gas production rises by around 40% between 2005 and 2050, and by a slightly higher 45% under the High estimate. It is only under the Low resource outcome that resource availability substantially limits growth in domestic production and use. In that case, gas production and use plateau around 2030 and are in decline by 2050. (MIT, 2011, 56)

Figure 8.2 sums up these long-range forecasts, but it is entirely plausible that by the 2030s or 2040s, absolute volumes and the market shares of natural gas might be somewhat below recent expectations, both globally and for the United States. On the other hand, I do not see any justification for going as far in the other direction as McNutt (2013): she suggested that the United States could deplete domestic gas rapidly and that the country could become dependent for its energy needs on China, the country with the largest technically recoverable shale gas resources. Whatever the future specifics, the two main uncertainties in decades ahead are the speed and extent of shale gas extraction in the United States and in half a dozen countries with major shale gas resources and the growth of LNG trade.

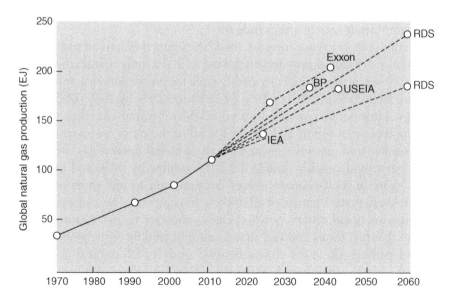

Figure 8.2   Long-range global gas production forecasts.

## 8.2   SHALE GAS PROSPECTS

Several critical matters that will decide the long-term future of shale gas have been either largely ignored or commonly glossed over, not only by media reporting but even by some of those shale gas enthusiasts whose professional understanding of oil and gas development should have made them to be more critical. Part of the problem is the fast progress of shale gas expansion. The shift toward horizontal drilling and hydraulic fracturing has been so rapid that it made even appraisals in leading energy publications instantly obsolete. Just one example to illustrate the point: on a paper in *Energy Policy* published in 2011, a group of experts concluded that LNG will become "the largest source of net U.S. imports by 2020" (Kumar et al., 2011, 4104)—while just two years later other experts were charting the US dominance of global LNG export market.

This surprising speed of American shale gas development resulted in no small expectations: dreams of potential benefits are, as befits a gaseous substance, properly inflated, and a common conclusion seems to be that shale gas will turn out to be a truly exceptional resource. Although nobody is repeating the infamous claim about nuclear energy in the 1950s (that it would be too cheap to meter), natural gas is predicted to remain inexpensive even as its domestic consumption rises and as intercontinental exports help to reverse America's balance of payments, to undercut the dominance of Russian exports to the European Union, to provide Asia with a cheaper alternative, and to assure America's strategic supremacy for decades to come. And domestically, the cheap fuel will attract not only petrochemical industries but also energize America's manufacturing renaissance and create large numbers of jobs (Dow Chemical, 2012).

Price Waterhouse Cooper concluded that shale gas is reshaping the US chemical industry (PWC, 2012). By 2013, the plans were for up to $100 billion of new petrochemical capacities, with investment by major US companies (Dow Chemical, ExxonMobil, Chevron) but nearly half of the total by foreign companies including South African Sasol, Saudi Arabian SABIC, and Taiwan's Formosa Plastics (Kaskey, 2013). American Chemistry Council estimated that between 2010 and 2020 new chemical capacities worth $71.7 billion will directly create 485,000 jobs and also lead to an additional $122 billion of indirect output and bring the total investment of $193 billion and 1.2 million new jobs (ACC, 2013). Inexpensive supply of methane and ethane would continue to attract new investment, but while building a typical modern petrochemical plant requires hundreds, or even thousands, of workers, operating plants will employ relatively few people. Consequently, future long-term job opportunities should not be exaggerated.

In any case, here is just a small sample of approbatory sentiments regarding expanding shale gas extraction by leading institutions and media sources (note all those superlative nouns and adjectives). As already noted, in June 2012, the International Energy Agency issued a report entitled *Golden Rules for a Golden Age of Gas* (IEA, 2012). In July 2012, *The Economist* reviewed the unconventional "bonanza" and saw this new source of gas "transforming the world's energy markets" (Wright, 2012). In February 2013, *Der Spiegel*, Germany's largest weekly newsmagazine, concluded that the US fracking has "the potential to shift the geopolitical balance in its favor" (Spiegel, 2013). In a few opening paragraphs of short news item in March 2013, CNBC, the leading US business TV channel, labeled the US energy developments as a power shift, an amazing turnaround, a renaissance, a miracle, a boom, and a bounty that "could shake up global order" (CNBC, 2013a, 1). Two weeks later, it raised a possibility of oil prices falling as low as $70/barrel as a result of these developments (CNBC, 2013a).

In October 2013, before a single new LNG export shipments of the US gas took place, *The American Oil & Gas Reporter* saw the US natural gas industry "positioned for dominant role in global LNG markets" (Weissman, 2013). And in the same month, Jaffe and Morse (2013) offered a much more expansive argument, claiming that

by providing ready alternatives to politicized energy supplies, the United States can use its influence to democratize global energy markets, much the way smartphone and social media technologies have ended the lock on information and communications by repressive governments and large multinational or state-run corporations,

and that this development will mean nothing less than "The end of OPEC." And as the conflict between Ukraine and Russia intensified during the spring of 2014, *Financial Times* was just one of many opinion makers suggesting a rethink of the US policy: "booming US oil and gas production has been portrayed as a bounty that will boost America's energy independence, but the crisis in Ukraine has cast it in a different light: as a strategic weapon to help allies overseas" (Jopson, 2014, 1).

All of these arguments, claims, conclusions, and suggestions should be challenged because most of them represent overenthusiastic reactions to what are undoubtedly important changes and because some of these deductions are plain cases of counterproductive wishful thinking. Most notably, I do not see either a swift demise of OPEC or the Ukrainian economy energized by massive US oil and gas exports. More importantly, except for Western Canada's shale gas extraction, no other country has embarked on exploration and extraction activities that would be even remotely resembling (even after adjusted for country size or economic level) the scale and intensity of the US shale fracking: clearly, that revolutionary example has not, so far, traveled well.

The two key questions about shale gas development that have yet to be satisfactorily answered are the extent of ultimately recoverable resources and the productivity decline of wells. In trying to answer the first question, Berman (2010) tried to discern some longer-term patterns by turning to Barnett shale, the first major development of its kind with cumulative experience based on some 8,000 horizontal wells. His findings surprised him:

…most reserve predictions based on hyperbolic production decline methods were too optimistic when compared with production performance. There is little correlation between initial production rates and ultimately recoverable reserves. Average well life is much shorter than predicted, and the volume of the commercially recoverable resource has been greatly over-estimated. (Berman, 2010, 1)

Berman's revised projection of ultimately recoverable reserves was 30% lower than his first estimate made just 2 years earlier. The main reason is that wells may not maintain the hyperbolic decline typical of the first months or years of their production: there may be sudden, catastrophic decreases, typically during the fourth of the fifth year of production but sometimes after just a year. Additional stimulation may temporarily boost the flow, but more often, it is followed by an even steeper output decline. As for an average well longevity, Berman argues that it does not make sense to expect at least 40 years of production instead of choosing an economic limit of about 57,000 m³/month (2 million cuft/month) as the threshold below which costs get higher than revenues based on the price of $3.5/1000 cuft, 25% royalty and average operating costs disclosed in the SEC filings.

These are critical differences: while the USGS put the ultimately recoverable Barnett shale gas at 736 Gm³, Berman has it at less than 255 Gm³ from nearly 12,000 wells, and he believes that additional 23,000 wells (with the land leased and wells drilled and completed at a cost of about $75 billion) would be needed to reach 736 Gm³. His two final surprising findings are the following: horizontal Barnett wells do not have significantly higher ultimate recovery than vertical wells—the difference is 31%, but the cost is 2.5 times higher—and performance of horizontal Barnett wells has not improved with time due to gains in experience and technical advances.

And a number of new resource appraisals—done with a much higher spatial resolution than used in the USEIA studies (blocks of a square mile that is 2.6 km$^2$ rather than by county, with some counties larger than 1000 km$^2$)—are coming up with much lower production forecasts (Patzek, Male, and Marder, 2013; Inman, 2014). While the USEIA sees natural gas production (largely driven by shale gas) growing until 2040, these new, much more conservative assessments see shale gas production from the four largest basins (Marcellus, Barnett, Fayetteville, and Haynesville) peaking already in 2020, with the 2030 output only about half of the USEIA's expectations and a steeper decline afterward. If true, the US energy outlook would become much dimmer in a matter of years. At the same, these new conservative assessments do not take into account any innovative exploration and recovery techniques: some of them will certainly make a significant difference during the decades to come.

General trajectory of gas production (and, similarly, oil) from horizontal wells stimulated by hydraulic fracturing is easily stated. A typical course begins with a steep hyperbolic decline for the first 2–3 years after the initial production peak, followed by a long exponential (constant annual percentage) tail decline until the production stops: estimated ultimate recovery is based on a 30-year lifespan, but it would be surprising if commercial life of most shale gas wells would be longer than 20 years. Initial flow rates from fractured shales are between 60,000 and 120,000 m$^3$/day, but the flows follow a pronounced hyperbolic decline whose magnitude differs among the major plays.

Sandrea's (2012) preliminary evaluation of production potential of mature US gas shale plays, based largely on data from Barnett and Fayetteville, showed how different these resources are from conventional gas. Shale gas recovery is characterized by unusually high annual field decline rates (ranging from 63% for Fayetteville to 86% for Haynesville), low ultimate recovery rates, and hence low recovery efficiencies averaging just 6.5% and ranging from just 4.7% for Haynesville, 5.8% for Barnett, and 10% for Fayetteville. In Pennsylvania's Marcellus shale, wells may produce more than 10 Mm$^3$ during the first year (average is nearly 6 Mm$^3$), less than 2 Mm$^3$ in the second year, and less than 1 Mm$^3$ in the third year (Harper and Kostelnik, 2009; King, 2014). In contrast, recovery rates range from 75 to 80% for conventional gas fields. Analysis of production potential showed that shale gas plays peaked at rates 2.6 times those of conventional deposits of the same size.

Perhaps the most representative findings of productivity decline in shale gas extraction came out of a comprehensive study of natural gas wells in five major shale gas basins (Barnett, Fayetteville, Woodford, Haynesville, and Eagle Ford) conducted by Schlumberger (Baihly et al., 2011). The study analyzed 1,885 representative wells (the largest number, 838, in Barnett shale) in a uniform manner that allowed for clear comparisons of performance starting on the date of first production and that produced the best possible estimates of ultimate recovery and hence allowed revealing assessments of long-term economic feasibility.

In Barnett shale, the maximum decline trend for wells drilled between 2003 and 2009 showed relatively small fluctuations around the following curve: 50% decline at the end of the first year, approximately 70% decline after 3 years, and 80% decline after 5 years. But the study concludes that Barnett shale is an exception as it had a flatter production decline trend, and hence, it would not serve as an analog for estimating productivity declines in other major shale plays. Fayetteville shale had a 1-year decline of about 60%, Woodford a bit over 60%, and Haynesville about 80% (Figure 8.3). Well costs ranged from $2.8 million in Fayetteville to $6.7 million in Woodford ($3 million in Barnett), operating costs (in $/Mcf) had a 3.5-fold span from 0.7 for Barnett to 2.5 in Haynesville, and ultimate recoveries per well were as low as 72 Mm$^3$ for Woodford wells drilled in 2008 to 172 Mm$^3$ for Haynesville wells drilled in 2009.

There has been nothing really surprising about the findings concerning decline rates (many less comprehensive earlier studies found similar rates), well costs, and ultimate recoveries, but the conclusions regarding the economic break-even price are surprising: for Barnett wells drilled in 2008 and 2009, they were only $3.70–3.74; for Fayetteville wells drilled in the same years, they were even lower at $3.20–3.65; but for Haynesville, they were $6.10–6.95, for Eagle Ford $6.10, and for Woodford $6.22–7.35. Barnett and Fayetteville could make money with gas at $4/Mcf at 10% discount rate and be profitable under low spot gas prices—but Haynesville and Eagle Ford and Woodford would need at least $6/Mcf.

But it is important to note that many operators hedge some or all of their production at higher than spot price values and that accumulating experience and technical innovation will tend to lower the future costs of extraction. At the same time, between 2010 and 2014, even as the extraction was rising and productivity per well was increasing in nearly

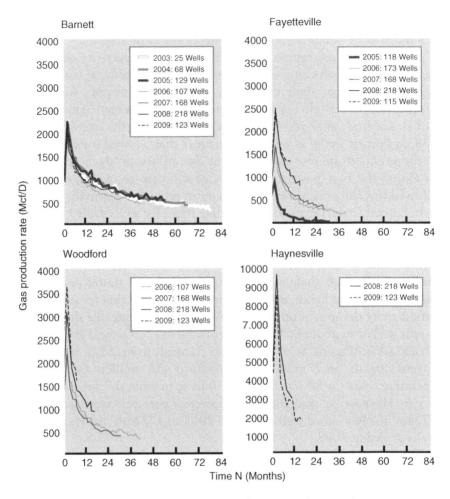

**Figure 8.3**    Decline of shale gas well output in the United States.

all shale gas formations, debt burden of America's shale oil and gas companies had nearly doubled, while revenues grew at just 5.6% during that period (Loder, 2014a). Sandrea (2014) also notes this reality—high capital expenditures nearly matching total revenues with net cash flow declining and debt rising—and also points out a relatively large number of substantial write-downs as many companies found the extraction much less attractive than they initially believed.

Is the US shale boom akin to a treadmill of capital spending, with dire consequences well known from other boom-and-bust cycles? In early 2014, 75 out of 97 oil and gas exploration and production companies rated by S&P were below investment grade (Loder, 2014a). Will bankruptcies of smaller and most overextended companies soon lead to more rational production or will be there a substantial investment decline? Because shale gas has become such an important part of the US energy supply so rapidly, we still should withhold any definite judgments regarding its eventual cumulative production and environmental impact.

For every superlative appraisal and endorsement cited earlier in this chapter, I could supply opinions ranging from much more subdued appraisals to calls for outright bans of shale gas production using hydraulic fracturing. In October 2013, Matthias Bichsel, projects and technology director at the Royal Dutch Shell, suggested that the US hydrocarbon industry has "overfracked and overdrilled" and reminded the enthusiasts that not all fields are created equal: Bakken, Eagle Ford, and Marcellus get all the attention, but the industry has not been able to achieve the same success in the Green River Basin in Wyoming and Utah (Hussain, 2013).

David Hughes, a Canadian geoscientist now at the Post Carbon Institute, concludes that "although the extraction of shale gas and tight oil will continue for a long time at some levels, production is likely to be below the exuberant forecasts from industry and government" (Hughes, 2013, 307). But some of these published expectations have not been excessive: the USEIA estimates that by 2030, shale gas will supply 7% of global natural gas production, hardly an excessive share. Similarly, the latest long-term price forecasts do not appear unrealistically low, assuming that by 2030 the gas for industrial consumer will be 85% more expensive than in 2012 and that residential consumers will pay about 30% more (USEIA, 2014i), an increase that might cover most, or all, higher costs of future shale gas extraction.

Environmentalists object to shale gas because its large-scale development and construction of pipeline and LNG infrastructure would be locking-in carbon emissions—some even claim at a level comparable to coal combustion (and, certainly, new gas-fired electricity-generating plants built recently will operate at least until 2040 to recoup the investment and make expected profit)—and making it more difficult to lower carbon emissions by adopting new reduction or capture techniques unless, of course, governments were to offer even greater incentives. As I explained in some detail (in Chapter 6), others object to any extensive development mainly because of the potential impact on freshwater resources.

On March 18, 2014, leaders of 16 environmental organizations sent a letter to President Obama asking him to stop any exports of LNG derived from shale gas because we are disturbed by your administration's support for hydraulic fracturing and, particularly, your plan to build liquefied natural gas export terminals along U.S. coastlines that would ship large amounts of fracked gas around the world. We call on you to reverse course on this plan and commit instead to keeping most of our nation's fossil fuel reserves in the ground. (McKibben et al., 2014, 1)

Writers of the letter also make a direct link between LNG from fracked gas and catastrophic climate change:

> Emerging and credible analyses now show that exported U.S. fracked gas is as harmful to the atmosphere as the combustion of coal overseas—if not worse. We believe that the implementation of a massive LNG export plan would lock in place infrastructure and economic dynamics that will make it almost impossible for the world to avoid catastrophic climate change… President Obama, exporting LNG is simply a bad idea in almost every way. We again implore you to shift course on this disastrous push to frack, liquefy, and export this climate-wrecking fossil fuel. (McKibben et al., 2014, 1–2)

Of course, such a stance would essentially eliminate any further combustion of all fossil fuels, an impossibility in a civilization where they now supply about 86% of all primary energy.

And factors other than productivity of shale gas wells and their environmental impacts will affect the extent of future recovery: these will include, above all, competition from other fossil fuels, renewable conversions, and nuclear electricity generation. Because of China's and India's demand surge, coal's share in global primary energy use has been actually rising since the beginning of the twenty-first century, and despite the intended goals of reducing coal combustion and consuming more natural gas, cost advantages of coal may make this substitution slower than envisaged. Hydraulic fracturing is, of course, also a major new source of crude oil: as a result, the United States became again not only the world's largest producer of natural gas but (when crude oil and natural gas liquids are combined) also the leading producer of liquid hydrocarbons in June 2014, the primacy it lost to the former USSR in 1975.

A new wave of nuclear power plant construction would reduce the demand for natural gas used for electricity generation, but given the parlous state of commercial fission in all Western countries and in Japan (stagnating, declining, on the way out, banned outright), this would matter only in Asia (China, South Korea, India), and even the greatest imaginable progress would be unlikely to displace large volumes of natural gas. Hence, the opposite consideration is much more pertinent: how much will cheap natural gas contribute to (once again) postponed renaissance of nuclear electricity generation, particularly in the United States, a phenomenon we have been promised since the mid-1980s?

Already, the effect has been undeniable as applications for construction of 12 new units have been either withdrawn or were suspended (USNRC [US Nuclear Regulatory Commission], 2014). Competition by wind and solar electricity generation could be much more important if judged simply by the total output of these renewable conversions—but, in reality, expansion of those intermittent renewable capacities would almost certainly increase the need for backup and peak demand gas-fueled generation during calm and cloudy days and (in the absence of mass-scale electricity storage) also for nighttime supply.

Global future of shale gas is even more uncertain. As already noted, even after nearly a decade after the US extraction began to take off, only Canada has become engaged in extensive shale gas production, while there is no sign of any similarly aggressive development of this new resource either in Europe or in Asia. In Poland, one of a few EU countries with interest in developing shale gas, detailed reevaluation of resources found them to be only a fraction of the estimates in EIA/ARI report (Kuuskraa, 2013). Looking ahead, it is not unreasonable to think that the combination of lower-than-expected endowment, environmentally based opposition, lack of readily available drilling, and well completion skills as well as the availability of requisite fracking capacities, regional shortages of water, and surprisingly high cost of extraction from deep, thin, and complex deposits will result in only slow rates of shale gas developments, not in replications of Barnett or Marcellus experience.

Yet another uncertainty comes with China's determination to use its abundant coal in a cleaner manner. China's first coal-to-gas project (Hexigen in Nei Monggol) began deliveries to Beijing at the end of 2013, starting with first 1.33 $Gm^3$/year and eventually rising to 4 $Gm^3$/year; in 2013, construction also began on three other projects, each with annual capacity of 4 $Gm^3$; and preparations started for five additional plants: if all completed, they would deliver annually 36 $Gm^3$ of gas (China News, 2013). If these projects were to prove successful and economically acceptable, the coal-rich country may emphasize coal-to-gas conversion rather than shale gas development, a combination of techniques where it has no deep technical expertise.

## 8.3   GLOBAL LNG

LNG has been the fastest-growing segment of global gas supply, but, once again, it would be unwise to overlook those factors that might moderate its further expansion. Perhaps the most important general consideration is the cost of new projects: there is no prospect for cheap LNG. Actually, the latest plants require considerably higher capital expenditures than the facilities built during the first years of the twenty-first century: upstream capital cost index of LNG developments has more than doubled between 2003 and 2013, and this rise has not been satisfactorily explained (IHS CERA, 2014; Songhurst, 2014). Typical rates during the 2000–2005 period were between $200 and 500/t of installed capacity. In contrast, even the latest low-cost projects (some involving conversion of regasification sites to liquefaction plants) range between $600 and 800/t of capacity; higher-cost projects (in Australia, Angola, and Papua New Guinea) range from $1,000 to 1,800; Norwegian Snøhvit (on a small Arctic island) cost $2,000/t (Figure 8.4); and among Australia's dozen new projects, one is above $2,000/t and five are above $3,000/t.

This means that a new high-cost LNG plant producing 5 Mt/year will cost at least $5 billion, development of the requisite gas supply will run (at $400/t of capacity) $2 billion, five LNG tankers ($220 million each) will add $1.1 billion, and regasification facility will require about $600

Figure 8.4   Snøhvit LNG plant. Reproduced with permission from statoil.com.

million, for the total cost of $8.7 billion. With construction delays, that can easily increase to $10 billion. That is why conversion of receiving terminals (so-called brownfield projects including Sabine Pass, Freeport, Lake Charles, Cove Point, and Cameron in the United States) is so appealing: converting the US regasification facilities to LNG plants would cut capital expenditures by at least 10%. Break-even price for most new LNG projects (be they in the United States, Africa, or Australia and based on conventional or nonconventional gas) is $10–11/million Btu (Fesharaki, 2013), and it might be possible to deliver the US shale gas (priced at $4/MMBtu) at less than $8/million Btu to Europe and at less than $10/million Btu to Asia (CB&I (Chicago Bridge and Iron Company), 2011).

So what is next? Continuing demand for LNG in Asia in general, and in East Asia in particular, is easy to predict. Natural gas is still only 10% of the continent's primary energy supply (only about 40% of the global mean) as established importers (Japan, South Korea, Taiwan) are being joined by China (first imports in 2006 to Guangdong), India (planning at least 10 new LNG terminals), Vietnam, and the Philippines. China has still a long way to go before surpassing Japan as the world's largest LNG importer (it bought about 25 Gm$^3$ in 2013 compared to Japan's 113 Gm$^3$), but the coming demand is huge, expected to rise to 315 Gm$^3$ in 2019, an increase of 90% above the 2013 volume, in order to displace more coal in household heating and city-based electricity generation and hence reduce excessive levels of urban air pollution.

As a result, Asian LNG imports could easily rise by 50% between 2010 and 2020—but the actual aggregate demand will depend on the eventual course of Japan's energy policy (how many nuclear stations will be eventually restarted?), South Korea's economic growth and expansion of nuclear capacities, and China's success in boosting domestic production and the country's extent of additional imports from Central Asia and Russia. The greatest import potential elsewhere in Asia is obviously in India, and its volume will be determined by the success of the new BJP government installed in 2014. Technical advances will result in a growing number of floating LNG plants designed to harness stranded gas (Figure 8.5). Self-propelled offshore liquefied natural gas production vessels can do it all: pretreat the gas, liquefy it, store it onboard, and off-load it to LNG tankers (Peck and van der Velde, 2013). Vessels by Flex LNG have annual production capacity of 1.7 Mt, while other developers are planning ships with capacities of 1.6–2.5 Mt. Floating regasification with permanently moored vessels, either offshore or in a port and ship-to-ship transfer of LNG, will also become more common.

Figure 8.5    Floating LNG plant. Reproduced with permission from Photographic Services, Shell International Limited.

As for individual LNG exporters, fairly predictable trends include both expected shifts and surprising changes. Continuing development of huge offshore natural gas reserves in the Indian Ocean off Western Australia—now mainly by two projects, Pluto by Australia's Woodside costing about $11 billion and Gorgon, a joint venture by Chevron, Exxon, and Shell, costing more than $40 billion—should make the country, now the world's fifth largest LNG exporter and major supplier of East Asia, second only to Qatar by 2020. In contrast, LNG imports by hydrocarbon-rich countries represent a shift that was unexpected even 5 years ago. In 2009, Kuwait was the first Gulf country to import LNG for an offshore storage to cover its summertime peak demand; it was followed by Dubai in 2010 (also for an offshore storage, in addition to pipeline imports from Qatar) and (as already noted) by Malaysia in 2013.

Dubai's profligacy is particularly astonishing. As if the world's tallest building (Burj Khalīfa) and the giant Dubai Mall shopping center were not enough, Dubai Holding (owned by the state's ruler Sheikh Mohammed bin Rashid Al Maktoum) announced in July 2014 that it is planning to build the Mall of the World, a temperature-controlled city under a giant glass dome with the world's largest mall and an indoor park, hotels, health resorts, and theaters that would occupy 4.45 Mm² and include 7 km long promenades (Dubai Holding, 2014). The proponents hope to provide "pleasant temperature-controlled environments during the summer months," making it "an attractive destination all year long" for anticipated more than 180 million annual visitors. In aggregate, Middle East's growing dependence on natural gas means that by 2025 the region's LNG import needs may approach 15 Mt/year, divided among Kuwait, UAE, and Saudi Arabia (Fesharaki, 2013).

Qatar's role in all of this is as crucial as it is uncertain. Politically, it is a curious amalgam: a supporter of Egypt's Islamist Muslim Brotherhood that has financed not only Al Jazeera but also jihadist groups in Syria's civil war and maintains links with the Taliban while hosting a large US air base at Doha. How long lasting will be this peculiar arrangement in the region that has already seen too many political upheavals? Moreover, the world's largest exporter of LNG has also seen its domestic natural gas consumption rising rapidly: it had nearly tripled between 2003 and 2012 (with most of the gas going for electricity generation and water desalinization), but the overall gas extraction had more than quadrupled. There is no imminent supply problem, but in 2005, the government decided to place moratorium on further development of the North Dome until 2014.

If Qatari shipments level off, how much can be supplied by North America's shale gas expansion? Canada's only LNG facility is Canaport, a receiving site in Saint John, New Brunswick, for imported gas from Trinidad and Tobago and Qatar—but there are plans for exports of British Columbia and Alberta natural gas piped across the Rocky Mountains to the Pacific coast with LNG terminals in Kitimat and Prince Rupert (LNG Canada, 2014). Both the timing and eventual magnitude of these shipments from the West Coast to Asia are highly uncertain, as even the requisite pipelines face a great deal of opposition by environmentalists and by Canada's aboriginals who are asserting their legal claims for land, and even for offshore waters, in their determination to stop any new development.

In any case, plans for the US LNG exports are much bolder. Expected dates for start-ups of already approved US export projects are Sabine Pass, LA, and Elba Island, GA, in 2016; Freeport, LA, Cove Point, MD, and Sabine Pass (Phase II) in 2017; and Lake Charles, LA, and Sabine Pass (Phase III) in 2018, and if all the pending proposals were to be approved, America's future LNG exports could be close to 50 Mt LNG by 2025 and eventually approach 200 Mt/year (compared to

Qatar's 77 Mt in 2013). But how many US projects will be eventually approved and how many of those will actually go ahead? While new brownfield LNG plants in the United States may have acceptable costs, it is far from certain that, as Weissman (2013) and others have claimed, the country is getting positioned for the dominant role in global LNG markets. Reaching that goal would depend on continued low or only mildly increased prices, but it does not strain credulity to think that the combination of growing domestic demand (caused by expanded gas-based electricity generation and new petrochemical capacities) and rising LNG exports could bring substantially higher prices (Ratner et al., 2013).

Could this unwelcome trend be hastened by concluding an excessive number of export agreements (currently there are nearly 40 applications to sell LNG abroad) that would reduce the number of new petrochemical projects whose construction rests on cheap natural gas? Dow Chemical thinks so, and the company has argued for restricted exports in order to protect domestic supplies (Helman, 2012). If most of the contemplated projects were to go ahead, the United States would become not only a major exporter but perhaps even the world's top seller of natural gas—but such a forecast seems to ignore the likely moves of other major gas exporters: becoming the top exporter of LNG is a goal that does not depend on the United States alone because whatever the United States does to advance that development could be affected, even negated, by the moves of other key global LNG players.

The largest LNG market in Asia (with Japan, China, South Korea, and Taiwan as the largest buyers) is now served by exports from Australia, Indonesia, Malaysia, and Qatar and by pipeline exports to China from Turkmenistan, Uzbekistan, and Kazakhstan. As already noted, the agreement reached in 2014 to supply China for 30 years with Siberian natural gas was a true strategic shift: annual volume of that contract is nearly as high as China's total (pipeline and LNG) gas imports in 2012, yet its price is lower than Gazprom's sales to Europe. Before this agreement was concluded, its volume could have been counted as a potential need to be filled by imported LNG, and if another similar agreement will be signed in 5 or 10 years, the volume of LNG imports China will seek from across the Pacific would decline further.

And is it reasonable to expect that Russia's Gazprom, the world's largest exporter of natural gas, would continue to insist on high long-term contract prices as the United States starts shipping significant volumes of LNG into Europe's ports? And what is Russia's long-term role in LNG trade? So far, Gazprom has been only on the fringes of global LNG trade with its Sakhalin project (gas deliveries to Japan started in March 2009), but the country has enormous Siberian gas reserves whose transportation to East Asian markets could be eased by Arctic warming. Would Gazprom simply stand aside and, after developing its Sakhalin reserves for exports to Japan and South Korea, concede the LNG market to others and concentrate only on pipeline shipments?

Hardly so, in 2012 LNG tanker *Ob River* charted by Gazprom completed the world's shipment via the Northern Sea Route, starting in Hammerfest, Norway, on November 7 and arriving at Tobata, Japan, on December 5 (Gazprom, 2012). Ice-breaking LNG tankers could access new resources of Arctic gas, already appraised to supply at least 30–40 Mt/year. And Russia's other large gas producer, Novatek, is developing (with state subsidies and together with France's Total and China's CNPC) a large LNG project on the Arctic's Yamal peninsula (South Tambeyskoye Field) that is valued at $27 billion and that would help to double Russia's share of global LNG market by 2020 to nearly 10% (Novatek, 2014a).

And would Qatar, the world's leading LNG seller and the owner of the largest and most modern tanker fleet, just stand by and watch as the tide of US exports reduces its European and Asian market shares? Would Australia, with its rich reserves and established Asian markets, abstain from competing for shipments to China and Japan? And how rapidly could large US LNG exports going to Europe change the EU's supply that has been dominated by Russia, Norway, the Netherlands, and Qatar? Would Gazprom, with its enormous sunk cost in long-distance pipelines from Western Siberia to the EU, just continue to insist on expensive and inflexible long-term contracts or would it try to undercut the US LNG exports to European ports?

And, ultimately, there is Iran, the country with the world's largest (and overwhelmingly undeveloped) conventional natural gas reserves. Eventually, Iran will become a normal country, not a theocracy ruled by fundamentalist mullahs. Such fundamental breaks seem always unthinkable, but they are bound to happen (just recall rapid transformations of power in the USSR, Egypt, or Myanmar), and if the country's hydrocarbon resources were to be developed with the benefit of the best foreign knowledge and technical assistance, its natural gas sales could easily surpass Qatar in total output and vie with Russia for global export primacy.

And given the widespread distribution of shales, almost every one of today's major LNG producers and many of today's LNG buyers will have to decide how far (if at all) they will go with their development and what balance they will eventually strike between increased domestic production and imports. Not surprisingly, by the end of 2014, many potential Asian buyers of the US LNG appeared to be much less enthusiastic about the future of such imports, reselling their quotas and reducing their commitments. And while there is no doubt that the progress toward a truly global, LNG-enabled, natural gas market will continue, the cost of new large LNG projects and inherently more expensive and less convenient transportation and storage of the fuel make it unlikely that the worldwide gas market will be as flexible as its crude oil counterpart.

Contradictory claims and disputes regarding hydraulic fracturing and new LNG systems will not be resolved in a year or two. We should be able to form a much more solid appraisal before 2020: by that time, intensive hydraulic fracturing in the United States will have accumulated more than a decade of intensive operation in every major shale formation, and as tens of thousands of wells will have yielded most of their expected ultimate recovery, we should be able to make a much more confident appraisal of long-term shale gas prospects. Moreover, by that time, environmental problems associated with shale gas extraction will have been either effectively managed or they will become a substantial drag on further development, and a new, truly global, LNG infrastructure will, or will not, have largely taken shape.

## 8.4   UNCERTAIN FUTURES

A seemingly strong consensus is no reliable indicator of future developments: it may be quite correct, but it is much more likely that actual developments will depart significantly from the anticipated course because the prevailing appraisals will miss some major developments and even many more less obvious but cumulatively critical shifts and adjustments. Evolution of modern energy systems illustrates all of these developments, ranging from richly fulfilled expectations of electricity's transformational role in lighting to hugely exaggerated hopes for nuclear generation. Recent consensus about the global future of natural gas is certainly not as uncritically unrealistic as were the hopes for nuclear fission in the early 1970s (consensus saw it as the dominant, if not the sole, mode of electricity generation by the century's end), but it is also far from being a matter of subdued expectations.

The International Energy Agency went all the way, asserting that "natural gas is poised to enter a golden age" (IEA, 2012); General

Electric stayed away from adjectives and entitled its survey *The Age of Gas* because the fuel "is poised to capture a larger share of the world's energy demand" (Evans and Farina 2013, 3). And common descriptors encountered in papers, reports, and books dealing with natural gas have included repeatedly revolution, renaissance, boom, bonanza, and transformation. And the MIT natural gas study highlighted the role of natural gas as "the cost benchmark against which other clean power sources must compete to remove the marginal ton of $CO_2$" and as "a cost-effective bridge to a low-carbon future" (MIT, 2011, 2).

I will adhere to my steadfast refusal to engage in any long-term forecasting, but I will restate some basic contours of coming development before I review a long array of uncertainties whose eventual resolutions could all have, directly and indirectly, major consequences for global and national extraction and conversion of natural gas. Perhaps most importantly, and contrary to many overenthusiastic recent appraisals, rising extraction of natural gas still has a long way to go before the fuel becomes to rival its solid and liquid fossil counterparts. A key reality is worth stressing once more: global transition from coal and oil to natural gas has been proceeding at a slower pace than the two previous great transitions from coal to oil and from wood to coal.

After it reached 5% of the global fossil fuel market, crude oil needed 40 years to reach 25%, while natural gas needed about 55 years to reach the same level. And if the fuel shares are expressed in terms of total primary energy supply (including all primary electricity and commercial biofuels), then crude oil peaked at about 48% in 1973, but natural gas is yet to reach 25% of the global total. Given the scale of existing energy demand and the inevitability of its further growth, it is quite impossible that during the twenty-first century, natural gas could come to occupy such a dominant position in the global primary energy supply as wood did in the preindustrial era or as coal did until the middle of the twentieth century.

Indeed, it is highly unlikely that even by 2050, the share of natural gas in the world's energy use would reach the peak that was attained during the 1970s by crude oil. And in cumulative energy terms, during the twentieth century, natural gas was the distant third fossil fuel: coal supplied about 43% of all fossil energy, crude oil about 37%, and natural gas 20% (Smil, 2010a). Cumulative output during the second half of the twentieth century saw coal and oil switching places (coal at 36%, crude oil at 43%), but natural gas rose only to 23%. Again, it is unlikely that natural gas will supply, in aggregate, more energy during the first half of the twenty-first century than either coal or crude oil.

And if the first decade of the twenty-first century was a trendsetter, then all fossil energy sources will cost substantially more, both to develop new capacities and to maintain production of established projects at least at today's levels. The IEA concluded that by 2013, capital costs of energy production have more than doubled in real terms (to $1.6 trillion) when compared to the year 2000 (IEA, 2014), and obviously, such large investments lock in the composition of primary energy use and consequence of such usage. The IEA estimates that between 2014 and 2035, the total investment in energy supply will have to reach just over $40 trillion if the world is to meet the expected demand, with some 60% destined to maintain existing output and 40% to supply the rising requirements. The likelihood of meeting this need will be determined by many other interrelated factors.

Combination of long lead times needed to develop new fuel extraction capacities and to build requisite processing and transportation infrastructures, and of rising costs of those facilities requiring long-term financial commitments mean that there can be no production surprises. And the principal actors and their roles also change slowly: as IGU (International Gas Union) (2012, 5) rightly concluded, "the rigid, long-term nature and economic impact of international gas transactions not only require more government involvement on both the producer and consumer end—and in any other transit country—than other forms of international energy transaction, but may also stir up wider political interest."

Consequently, we have a fairly good understanding of what will take place during the next 2 or 3 years—with the usual proviso that even short-term confidence can be shattered, or at least much modified, by major international conflicts, by another substantial global economic downturn, or, least likely, by a crippling pandemic or an encounter with an asteroid. But forecasting even a decade ahead multiplies the uncertainties that have been made fundamentally more interrelated by the new nature of interdependent global economy and politics that has been reshaping the basic contours of post-1945 international order.

Here is a far from exhaustive list of factors that have a high probability of affecting (be it positively or negatively) the world's natural gas supply in near term (during the next 5–10 years). A large number of these factors concerns hydrocarbon resources and technical advances in their recovery and conversion. Extent of the US shale gas production (continuing rise, an early plateau, and unexpected decline) will make long-term prospects much clearer. Progress of shale gas extraction in other countries, above all in China (lifting of fracking bans; aggressive pursuit in several countries), would indicate if shale gas will remain largely a North American phenomenon or if it will become a global fuel. Decline of conventional oil reserves and the success in large-scale recovery of nonconventional oil will determine the supply or shortfall of liquid hydrocarbons that could be partially or largely substituted (via GTL conversions) by gas. Reduction of fuel demand due to continuous rise of conversion efficiencies, both in stationary applications and in transportation, would moderate demand for all hydrocarbon fuels.

Another set of factors include direct and indirect competition arising from the development of other energy sources and from environmental concerns. Will there be a widespread acceleration (along the German model) of wind and solar generation (based on new, larger, superior turbines and on appreciably high PV panel efficiencies) that would be substantial enough to affect demand for natural gas used to produce electricity? Or will these renewable sources be boosted by diffusion of new promising methods for medium- to large-scale electricity storage that would enable to bank cheap generation during sunny and windy spells and use it at times of peak demand and during night? Will we see a relative success, or yet another failure, in commercializing hydrogen-powered vehicles? Japanese automakers are now particularly eager to develop a new market and are willing to take a long view and incur substantial initial losses (Ohnsman, 2013).

Will evolution and adoption of environmentally benign hydraulic fracturing, combined with success in limiting water contamination and fugitive methane missions in the US shale gas extraction, make shale gas a highly appealing (even if, in many countries, relatively expensive) choice? Will China, the world's largest consumer of coal with commensurately severe air pollution problems, be forced to accelerate its transition from coal to natural gas, or will its progress be relatively slow? Even more importantly, will a sudden reacceleration of global temperature rise—after a relatively lengthy pause, with the global average up by only 0.04°C/decade since 1998 (Tollefson, 2014; Figure 8.6)—lead to an unprecedented quest for less carbon-intensive fuel?

Political uncertainties may be no less important. Will Russia pursue a path of economic stability and cooperative foreign relations or will its behavior remain problematic, affecting its relations and trade with the EU and its position as a global supplier of energy? What will be the course pursued by India under new management? If steered right, the country could begin to rival China in its demand for all kinds of resources. What will the continuing destabilization and fracturing of the Middle East (I cannot assume the opposite, a sudden resolution of

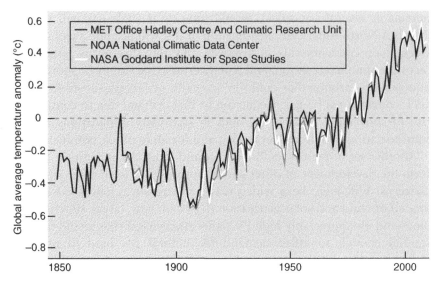

**Figure 8.6** Global warming pause.

multifaceted problems besetting the region) do to the global supply of hydrocarbons.

As greater uncertainties arise and new possibilities open up in more distant future, an analogical list looking at resource, technical, economic, environmental, and political factors that might become important in long term, namely, between 2025 and 2050, could be much longer, even after leaving out speculations about new technical breakthroughs (such as inexpensive small-scale GLT and easy and safe recovery of methane hydrates). In a few decades, many national trajectories, bearing on the global market for natural gas, should become clearer, and here are just a few terse paragraphs, with items given in no particular order as it is impossible to attach relative impact weights and occurrence probabilities to individual changes and events.

For the United States, still the world's most important economy, will it be an arrival of a new equilibrium as a more thoughtful, less wasteful power concerned about quality (of life, of civil society, of the global environment) rather than obsessed with growth and quantity? For Russia will it be economic success or failure (made worse by demographic and health problems)? And its evolving role in the global economy (with possibilities ranging from moving close to the EU to an uneasy state reminiscent of the Cold War era) will determine the extent of its remaining vast resources reaching foreign markets. Russian failures could be compounded or greatly compensated for by changes in Iran. The end of Iran's theocracy and the country's successful transition to a modern society would strengthen long-term stability of global natural gas supply. And will China's continuing rise improve or further worsen its relations with great powers and neighbors, or will it weaken and leave the country still far short of really modern society? And will the EU carry on, implode, or become largely irrelevant as mass migrations and rise of Asia further reduce its prospects?

What will be the ultimate fate of nuclear electricity generation: will it prove definitely too costly and too risky, or, finally, will it see a long-ago promised renaissance? Of course, should we be able to master reliable, affordable and environmentally acceptable recovery of methane hydrates, neither the fates of nuclear industry nor declining conventional oil production would matter that much: commercial hydrates would amount to a truly revolutionary departure. And progress in the creation of hydrogen economy (be it based on inexpensive solar achieved through new technical breakthroughs or, to be really frivolous, by success in fusion) would change the entire global calculus of energy supply and demand.

Natural gas is an excellent fossil fuel whose many inherent advantages guarantee its increasing use; much like all other energy sources, its greater use also has its share of drawbacks and complications, be they technical, financial, or environmental. And in the game played by the rules of dynamic global energy market, it is still, and in foreseeable future it will remain, nothing more than one of several key cards. This means that a more accurate characterization of the coming decades of changes in global fossil fuel composition would be not the age of gas but the era of rising natural gas importance. And the nature of the complex energy system subjected to these changes is such that during the first half of the twenty-first century, it will be, inevitably, an evolutionary rather than revolutionary shift: definitely not an age of brass but also, most likely, not a golden age.

# References

Abrams, L. 2014. Fracking's untold health threat: How toxic contamination is destroying lives. *Salon*, August 2, 2014. http://www.salon.com/2014/08/02/frackings_untold_health_threat_how_toxic_contamination_is_destroying_lives/ (accessed February 20, 2015).

ACC (American Chemistry Council). 2013. Shale gas, competitiveness, and new US chemical industry investment: An analysis based on announced projects. http://chemistrytoenergy.com/shale-study (accessed February 20, 2015).

Acton, A. et al. 2013. LNG incident identification—Updated compilation and analysis by the International Group of LNG Importers (GIIGNL). http://www.gastechnology.org/Training/Documents/LNG17-proceedings/05_02-Anthony-Acton-Presentation.pdf (accessed February 20, 2015).

Adamchak, F. and A. Adede. 2013. LNG as marine fuel. http://www.gastechnology.org/Training/Documents/LNG17-proceedings/Transport-11-Fred-Adamchak-Presentation.pdf (accessed February 20, 2015).

Adshead, A.M. 1992. Salt and Civilization. London: Palgrave Macmillan.

AEGPL (European LPG Association). 2014. AEGPL response to Commission's consultation. http://ec.europa.eu/reducing_co2_emissions_from_cars/doc_contrib/aegpl_en.pdf (accessed February 20, 2015).

AGA (American Gas Association). 2006. The role of energy pipelines and research in the United States. http://chemistrytoenergy.com/sites/chemistrytoenergy.com/files/shale-gas-full-study.pdf (accessed February 20, 2015).

Allen, D.T. et al. 2013. Measurements of methane emissions at natural gas production sites in the United States. Proceedings of the National Academy of Sciences of the United States of America 110:17768–17773.

Alliston, C., Banach, J. and J. Dzatko. 2002. Liquid skin. World Pipelines 2002(6):55–58.

Almqvist, E. 2003. History of Industrial Gases. Berlin: Springer.

Alstom. 2007. The world's first industrial gas turbine set—GT Neuchâtel. https://www.asme.org/wwwasmeorg/media/ResourceFiles/AboutASME/Who%20We%20Are/Engineering%20History/Landmarks/135-Neuchatel-Gas-Turbine.pdf (accessed February 20, 2015).

Alvarez, R.A. et al. 2012. Greater focus on methane leakage from natural gas infrastructure. Proceedings of the National Academy of Sciences of the United States of America 109:6435–6440.

Arthur, D. and D. Cornue. 2010. Technologies reduce pad size, waste. *The American Oil & Gas Reporter*, August 2010. http://www.all-llc.com/publicdownloads/AOGR-0810ALLConsulting.pdf (accessed February 20, 2015).

Ausubel, J. 2003. Decarbonization: The next 100 years. Lecture at the 50th Anniversary Symposium of the Geology Foundation, Jackson School of Geosciences, University of Texas, Austin, TX, April 25, 2003. http://phe.rockefeller.edu/AustinDecarbonization/AustinDecarbonization.pdf (accessed February 20, 2015).

Bai, G. and Y. Xu. 2014. Giant fields retain dominance in reserves growth. Oil & Gas Journal 112(2):44–51.

Baihly, J. et al. 2011. Study assesses shale decline rates. *The American Oil & Gas Reporter* May 2011:114–121.

Balling, L., Termuehlen, H. and R. Baumgartner. 2002. Forty years of combined cycle power plants. *ASME Power Division Special Section*, October 2001:7–30. http://www.energy-tech.com/uploads/17/0210_ASME.pdf (accessed March 11, 2015).

Bamberger, M. and R. Oswald. 2014. The Real Cost of Fracking: How America's Shale Gas Boom Is Threatening Our Families, Pets, and Food. Boston: Beacon Press.

Beerling, D. et al. 2009. Methane and $CH_4$-related greenhouse effect over the past 400 million years. American Journal of Science 309:97–113.

Begos, K. and J. Fahey. 2014. AP IMPACT: Deadly side effect to fracking boom. *AP*, May 5, 2014. http://bigstory.ap.org/article/ap-impact-deadly-side-effect-fracking-boom-0 (accessed February 20, 2015).

Begos, K. and J. Peltz. 2013. Anti-fracking celebrities, such as Yoko Ono, Mark Ruffalo and others, put 'fractivism' in the spotlight. http://www.huffingtonpost.com/2013/03/05/anti-fracking-celebrities-yoko-ono-ruffalo_n_2812726.html (accessed February 20, 2015).

Belvedere, M.J. 2014. Obama, Congress should have listened to me: Pickens. CNBC.com, July 7, 2014. http://www.cnbc.com/id/101815722 (accessed February 20, 2015).

Berman, A. 2010. Lessons from the Barnett Shale suggest caution in other shale plays. *First Enercast Financial*, March 29, 2010. http://www.firstenercastfinancial.com/commentary/?cont=3193 (accessed February 20, 2015).

BIA (Brick Industry Association). 2014. Manufacturing of Brick. http://www.firstenercastfinancial.com/commentary/?cont=3193 (accessed February 20, 2015).

Biocides Panel. 2013. Biocide Active Ingredients and Product Registration Status in Hydraulic Fracturing. Arlington: American Chemistry Council. http://www.aapco.org/meetings/minutes/2013/apr22/att8_purdy_acc_fracking.pdf (accessed February 20, 2015).

Bogislaw, G. 1991. Die Treibstoffversorgung durch Kohlehydrierung in Deutschland von 1933 bis 1945, unter besonderer Berücksichtigung wirtschafts- und energiepolitischer Einflüsse. Köln: Müller Botermann.

Boone Pickens. 2014. Boone Pickens. His Life. His Legacy. http://www.boonepickens.com/ (accessed February 20, 2015).

Boswell, R. 2009. Is gas hydrate energy within reach? Science 325:957–958.

Bowker, K.A. 2003. Recent developments of the Barnett Shale play, Fort Worth Basin. West Texas Geological Society Bulletin 42(6):4–11.

BP (British Petroleum). 2014a. *BP Energy Outlook 2035*. London: BP. http://www.bp. com/content/dam/bp/pdf/Energy-economics/Energy-Outlook/Energy_Outlook_2035_booklet.pdf (accessed February 20, 2015).

BP. 2014b. Statistical review of world energy. http://www.bp.com/content/dam/bp/pdf/Energy-economics/statistical-review-2014/BP-statistical-review-of-world-energy-2014-full-report.pdf (accessed February 20, 2015).

Brackett, W. 2008. A history and overview of the Barnett shale. http://www.bp.com/content/dam/bp/pdf/Energy-economics/Energy-Outlook/Energy_Outlook_2035_booklet.pdf (accessed February 20, 2015).

Brandt, A.R. et al. 2014. Methane leaks from North American natural gas system. Science 343:733–735.

Brantly, J.E. 1971. History of Oil Well Drilling. Houston: Gulf Publishing.

Brown, C. 2013. Gas-to-Liquid: A Viable Alternative to Oil-Derived Transport Fuels? Oxford: The Oxford Institute for Energy Studies. http://www.oxfordenergy.org/wpcms/wp-content/uploads/2013/05/WPM-50.pdf (accessed February 20, 2015).

BRS (Barry Rogliano Salles). 2014. 2014 annual review: Shipping and shipbuilding markets. http://www.brsbrokers.com/review_archives.php (accessed February 20, 2015).

Brune, A. 2010. Methanogenesis in the digestive tracts of insects. In: Handbook of Hydrocarbon and Lipid Microbiology, K.N. Timmins, ed., Heidelberg: Springer, pp. 707–728.

Burel, F., Taccani, R. and N. Zuliani. 2013. Improving sustainability of maritime transport through utilization of liquefied natural gas (LNG) for propulsion. Energy 57:412–420.

Burnham, A. et al. 2011a. Modeling the relative GHG emissions of conventional and shale gas production. Environmental Science & Technology 45:10757–10764.

Burnham, A. et al. 2011b. Life-cycle greenhouse gas emissions of shale gas, natural gas, coal, and petroleum. Environmental Science & Technology 46:619–627.

Buryakovsky, L. et al. 2005. Geology and Geochemistry of Oil and Gas. Amsterdam: Elsevier.

Cabot, G.L. 1915. Means for Handling and Transporting Liquid Gas. US Patent 1,140,250, May 18, 1915. Washington, DC: USPTO.

CAPP (Canadian Association of Petroleum Producers). 2014. Canadian association of petroleum producers. http://www.capp.ca/Pages/default.aspx (accessed February 20, 2015).

Castaneda, C.J. 2004. Natural gas, history of. In: Encyclopedia of Energy, C. Cleveland et al. eds., San Diego: Elsevier, Vol. 1, pp. 207–218.

Castle, W.F. 2007. Fifty-years' development of cryogenic liquefaction processes. In: Cryogenic Engineering, K.D. Timmerhaus and R.P. Reed, eds., New York: Springer, pp. 146–160.

Cathles, L.M. 2012. Comments on Pétron et al's (2012) inference on methane emissions from Denver-Julesburg Basin from air measurements at the Boulder Atmospheric Observatory tower. http://www.geo.cornell.edu/eas/PeoplePlaces/Faculty/cathles/Gas%20Blog%20PDFs/0.3%20Comments%20on%20Petron%20et%20al.pdf (accessed February 20, 2015).

Caulton, D.R. et al. 2014. Toward a better understanding and quantification of methane emissions from shale gas development. Proceedings of the National Academy of Sciences 111:6237–6242.

CB&I (Chicago Bridge and Iron Company). 2011. Current state & outlook for the LNG industry. http://www.forum.rice.edu/wp-content/uploads/2011/06/RT_110909_Humphries.pdf (accessed February 20, 2015).

CDC (Centers for Disease Control). 2013. Summary health statistics for the U.S. population: National Health Interview Survey, 2012. http://www.cdc.gov/nchs/data/series/sr_10/sr10_259.pdf (accessed February 20, 2015).

CDC. 2014. Accidents or unintentional injuries. http://www.cdc.gov/nchs/fastats/accidental-injury.htm (accessed February 20, 2015).

CDIAC. 2014. Carbon dioxide information analysis center. http://cdiac.ornl.gov/# (accessed February 20, 2015).

Center for Energy. 2014. Where are gas hydrates found? http://www.centreforenergy.com/AboutEnergy/ONG/GasHydrates/Overview.asp?page=3 (accessed February 20, 2015).

CEPA (Canadian Energy Pipeline Association). 2014. History of pipelines. http://www.cepa.com/about-pipelines/history-of-pipelines (accessed February 20, 2015).

CGA (Canadian Gas Association). 2014. Gas stats. http://www.cga.ca/resources/gas-stats/ (accessed February 20, 2015).

Chandler, D. and A.D. Lacey. 1949. The Rise of the Gas Industry in Britain. London: British Gas Council.

Chang, J. and J. Strahl. 2012. Shale gas in China: Hype and reality. http://pesd.stanford.edu/events/energy_working_group_shale_gas_in_china_hype_and_reality/ (accessed February 20, 2015).

Chang, A., Pashikanti, K. and Y. Liu. 2012. Refinery Engineering: Integrated Process Modeling and Optimization. Weinheim: Wiley-VCH.

Chessa, B. et al. 2009. Revealing the history of sheep domestication using retrovirus integrations. Science 324:531–536.

Chevron. 2014. Gas-to-liquids. http://www.chevron.com/deliveringenergy/gastoliquids/ (accessed February 20, 2015).

China News. 2013. China's first coal-to-gas project ready. China News, October 31, 2013. http://news.xinhuanet.com/english/china/2013-10/31/c_132848573.htm (accessed February 20, 2015).

China.org. 2014. West-East gas pipeline project. http://www.china.org.cn/english/features/Gas-Pipeline/37313.htm (accessed February 20, 2015).

Cho, J.H. et al. 2005. Large LNG carrier poses economic advantages, technical challenges. Oil & Gas Journal's LNG Observer 2(4):17–23. http://fred.caer.uky.edu/library/oilgasjnrl/2005_Oct_03_OilGasJournal.pdf

Christensen, T.R. 2014. Understand Arctic methane variability. Nature 509:279–281.

CIA. 1981. USSR-Western Europe: Implications of the Siberia-to-Europe Gas Pipeline. Washington, DC: CIA.

CIA. 2014. The world factbook. https://www.cia.gov/library/publications/the-world-factbook/rankorder/2242rank.html (accessed February 20, 2015).

Climate Registry. 2013. The climate registry's 2013 default emission factors. http://www.theclimateregistry.org/downloads/2013/01/2013-Climate-Registry-Default-Emissions-Factors.pdf (accessed February 20, 2015).

Clingendael. 2009. Crossing borders in European Gas Networks: The missing links. http://clingendael.info/publications/2009/20090900_ciep_paper_gas_networks.pdf (accessed February 20, 2015).

CNBC. 2013a. How the US oil and gas boom could shake up global order. CNBC.com, April 1, 2013. http://investigations.nbcnews.com/_news/2013/04/01/17519026-how-the-us-oil-gas-boom-could-shake-up-global-order?lite (accessed February 20, 2015).

CNBC. 2013b. Power shift: Energy boom dawning in America. CNBC.com, March 18, 2013. http://m.cnbc.com/us_news/100563497 (accessed February 20, 2015).

CNPC (China National Petroleum Corporation). 2014. Efficient development of the large tight sandstone gas field in Sulige. http://riped.cnpc.com.cn/en/press/publications/brochure/PageAssets/Images/pdf/16-Efficient%20Development%20of%20the%20 Large%20Tight%20Sandstone%20Gas%20Field%20in%20Sulige. pdf?COLLCC=1914310845& (accessed February 20, 2015).

COGEN Europe. 2014. COGEN Europe The European Association for the promotion of cogeneration. http://www.cogeneurope.eu/ (accessed February 20, 2015).

Collett, T.S. and G.D. Ginsburg. 1998. Gas hydrates in the Messoyakha gas field of the West Siberian Basin—A re-examination of the geologic evidence. International Journal of Offshore and Polar Engineering 8(1):22–29.

Conrado, R.J. and R. Gonzalez. 2014. Envisioning the bioconversion of methane to liquid fuels. Science 343:621–623.

Corkhill, M. 1975. LNG carriers: The Ships and Their Market. London: Fairplay Publications.

Cornwell, W.K. et al. 2009. Plant traits and wood fates across the globe: Rotted, burned, or consumed? Global Change Biology 15:2431-2449.

Crump, E.L. 2000. Economic Impact Analysis for the Proposed Carbon Black Manufacturing NESHAP. Washington, DC: USEPA.

Czuppon, T.A. et al. 1992. Ammonia. In: Kirk-Othmer Encyclopedia of Chemical Technology, 4th edition, M. Howe-Grant, ed., New York: John Wiley & Sons, Inc., Vol. 2, pp. 653-655.

Daemen, J.J.K. 2004. Coal industry, history of. In: Encyclopedia of Energy, C. Cleveland et al. eds., San Diego: Elsevier, Vol. 1, pp. 457–473.

Dallimore, S.R. et al. 2002. Overview of gas hydrate research at the Mallik Field in the Mackenzie Delta, Northwest Territories, Canada. http://www.netl.doe.gov/kmd/cds/disk10/Dallimore.pdf (accessed February 20, 2015).

Daly, J.C.K. 2009. Analysis: The Gazprom-Ukraine dispute. Energy Daily, January 12, 2009. http://www.energy-daily.com/reports/Analysis_The_Gazprom-Ukraine_dispute_999.html (accessed February 20, 2015).

Davis, L.D. 1995. Rotary, Kelly, Swivel, Tongs, and Top Drive. Austin: Petroleum Extension Service.

De Beer, J., Worrell, E. and K. Blok. 1998. Future technologies for energy-efficient iron and steel making. Annual Review of Energy and the Environment 23:123–205.

de Klerk, A. 2012. Gas-to-liquids conversion. In: Natural Gas Conversion Technologies Workshop. Houston: US Department of Energy, January 13, 2012. http://www.arpa-e.energy.gov/sites/default/files/documents/files/De_Klerk_NatGas_Pres.pdf (accessed February 20, 2015).

Deutsche Bank Group. 2011. Comparing Life-Cycle Greenhouse Gas Emissions from Natural Gas and Coal. Frankfurt: Deutsche Bank. http://www.worldwatch.org/system/files/pdf/Natural_Gas_LCA_Update_082511.pdf (accessed February 20, 2015).

Devold, H. 2013. Oil and Gas Production Handbook. Oslo: ABB Oil and Gas. http://www04.abb.com/global/seitp/seitp202.nsf/0/f8414ee6c6813f5548257c14001f11f2/$file/Oil+and+gas+production+handbook.pdf (accessed February 20, 2015).

Dillon, W.P. et al. 1992. Gas hydrates in deep ocean sediments offshore southeastern United States: A future resource? In: USGS Research on Energy Resources, Geological Survey Circular 1074, L.M.H. Carter, ed., pp. 18–21.

Dlugokencky, E.J. et al. 1998. Continuing decline on the growth rate of the atmospheric methane burden. Nature 394:447–450.

Dow Chemical. 2012. Natural gas: Fueling an American manufacturing renaissance. http://www.dow.com/energy/pdf/Dow_Nat_Gas_0801121.pdf (accessed February 20, 2015).

DSM. 2014. DSM. http://www.dsm.com/corporate/home.html (accessed February 20, 2015).

Dubai Holding. 2014. Mohammed Bin Rashid launches Mall of the World, a temperature-controlled pedestrian city in Dubai. http://www.dubaiholding.com/media-centre/press-releases/2014/407-mohammed-bin-rashid-launches-mall-of-the-world-a-temperature-controlled-pedestrian-city-in-dubai (accessed February 20, 2015).

Dubois, M.K. et al. 2006. Hugoton Asset Management Project (HAMP): Hugoton Geomodel Final Report. http://www.kgs.ku.edu/PRS/publication/2007/OFR07_06/ (accessed February 20, 2015).

Dukes, J.S. 2003. Burning buried sunshine: Human consumption of ancient solar energy. Climatic Change 61:31–44.

EC (European Commission). 2014. Energy prices and cost report. http://ec.europa.eu/energy/doc/2030/20140122_swd_prices.pdf

Edwards, P.M. et al. 2014. High winter ozone pollution from carbonyl photolysis in oil and gas basin. Nature 514:351–354.

EEGA (East European Gas Analysis). 2014. Major gas pipelines of the former Soviet Union and capacity of export pipelines. http://www.eegas.com/fsu.htm (accessed February 20, 2015).

Ehrenberg, S.N., Nadeau, P.H. and Ø. Steen. 2009. Petroleum reservoir porosity versus depth: Influence of geological age. AAPG Bulletin 95:1281–1296.

Energy Price Index. 2013. Household energy price index for Europe. http://www.energypriceindex.com/wp-content/uploads/2013/12/HEPI_Press_Release_December-2013.pdf (accessed February 20, 2015).

Engelder, T. 2009. Marcellus. *Fort Worth Basin Oil & Gas Magazine* August 2009:18–22.

Engelder, T. and G.G. Lash. 2008. Marcellus Shale play's vast resource potential creating stir in Appalachia. *The American Oil & Gas Reporter*, May 2008. http://www.aogr.com/magazine/cover-story/marcellus-shale-plays-vast-resource-potential-creating-stir-in-appalachia (accessed February 20, 2015).

ESC (Energy Solutions Center). 2005. Natural gas ... A cool solution to the high cost of cooling. http://www.gasairconditioning.org/pdfs/tools/cooling_guide.pdf (accessed February 20, 2015).

Esrafili-Dizaji, B. et al. 2013. Great exploration targets in the Persian Gulf: the North Dome/South Pars Fields. *Finding Petroleum*, February 13, 2013. http://www.findingpetroleum.com/n/Great_exploration_targets_in_the_Persian_Gulf_the_North_DomeSouth_Pars_Fields/ab3518c5.aspx#ixzz38WrohvKR; http://www.findingpetroleum.com/n/Great_exploration_targets_in_the_Persian_Gulf_the_North_DomeSouth_Pars_Fields/ab3518c5.aspx (accessed February 20, 2015).

Evans, P.C. and M.F. Farina. 2013. The age of gas & the power of networks. http://www.ge.com/sites/default/files/GE_Age_of_Gas_Whitepaper_20131014v2.pdf (accessed February 20, 2015).

ExxonMobil. 2013. The Outlook for Energy: A View to 2040. Irving: ExxonMobil. http://corporate.exxonmobil.com/en/energy/energy-outlook (accessed March 11, 2015).

Eyl-Mazzega, M.-A. 2013. Gas Markets and Supplies from the Former Soviet Union Area. Paris: IEA.

FAO (Food and Agriculture Organization). 2011. Current World Fertilizer Trends and Outlook to 2015. Rome: FAO. ftp://ftp.fao.org/ag/agp/docs/cwfto15.pdf (accessed March 10, 2015).

Fay, J.A. 1980. Risks of LNG and LPG. Annual Review of Energy 5:89–105.

Fesharaki, F. 2013. Key challenges for the development of LNG in emerging markets: Medium and long term outlook. Presented at the 17th International Conference & Exhibition on Liquefied Natural Gas, Houston, TX, April 16–19, 2013. http://www.gastechnology.org/Training/Documents/LNG17-proceedings/09_02-F-Fesharaki-Presentation.pdf (accessed February 20, 2015).

Fisher, J.C. and R.H. Pry. 1971. A simple substitution model of technological change. Technological Forecasting and Social Change 3:75–88.

Foss. 2004. Natural gas industry, energy policy in. In: Encyclopedia of Energy, C. Cleveland et al. eds., San Diego: Elsevier, Vol. 4, pp. 219–233.

Fouquet, R. 2008. Heat, Power and Light: Revolutions in Energy Services. Cheltenham: Edward Elgar.

Fraas, A.G., Harrington, W. and R.D. Morgenstern. 2013. Cheaper Fuels for the Light-Duty Fleet: Opportunities and Barriers. Washington, DC: Resources for the Future. http://chemistrytoenergy.com/sites/chemistrytoenergy.com/files/shale-gas-full-study.pdf (accessed February 20, 2015).

FracFocus. 2014. What chemicals are used. http://fracfocus.org/chemical-use/what-chemicals-are-used (accessed February 20, 2015).

Fractracker Alliance. 2014. Ratio of water used to product: How do Ohio drillers compare? http://www.fractracker.org/2014/10/water-energy-nexus-in-ohio/ (accessed February 20, 2015).

Freise, J. 2011. The EROI of conventional Canadian natural gas production. Sustainability 3:2080–2104.

Gallagher, M. 2013. The future of natural gas as a transportation fuel. http://www.eia.gov/conference/2013/pdf/presentations/gallagher.pdf (accessed February 20, 2015).

Garbutt, D. 2004. Unconventional Gas. Houston: Schlumberger. http://www.slb.com/~/media/Files/industry_challenges/unconventional_gas/white_papers/uncongas_whitepaper_of03056.ashx (accessed February 20, 2015).

Gazprom. 2012. Gazprom successfully completes world's first LNG supply via Northern Sea Route. http://www.gazprom.com/press/news/2012/december/article150603/ (accessed February 20, 2015).

Gazprom. 2014. Share price graph. http://www.gazprom.com/investors/stock/stocks/ (accessed February 20, 2015).

GE (General Electric). 2014a. LM6000 & SPRINT aeroderivative gas turbine packages (36–64 MW). https://www.ge-distributedpower.com/products/power-generation/35-to-65mw/lm6000-sprint-series (accessed February 20, 2015).

GE. 2014b. 2014 performance specs. http://efficiency.gepower.com/pdf/GEA31167%209H_GTW_Reprint_R6_SPREAD (accessed February 20, 2015).

GE Power & Water. 2013. Fast, flexible power: Aeroderivative product and service solutions. https://www.ge-distributedpower.com/products/power-generation? (accessed February 20, 2015).

Geoscience Australia. 2012. Coal bed methane. http://www.australianminesatlas.gov.au/education/fact_sheets/coal_bed_methane.html (accessed February 20, 2015).

GET (Grand Energy Transition). 2014. Hefner on the transition to natural gas (CNG) use in the transportation sector. http://www.the-get.com/hefner_views/index.php?page=CNG (accessed February 20, 2015).

GGFR (Global Gas Flaring Reduction). 2013. Global gas flaring reduction partnership. http://web.worldbank.org/WBSITE/EXTERNAL/TOPICS/EXTOGMC/EXTGGFR/0,,menuPK:578075~pagePK:64168427~piPK:64168435~theSitePK:578069,00.html (accessed February 20, 2015).

GGFRP (Global Gas Flaring Reduction Partnership). 2014. Initiative to reduce global gas flaring. http://www.worldbank.org/en/news/feature/2014/09/22/initiative-to-reduce-global-gas-flaring (accessed March 10, 2015).

Ghosh, A.K. 2011. Fundamentals of paper drying—Theory and application from industrial perspective. http://cdn.intechopen.com/pdfs-wm/19429.pdf (accessed February 20, 2015).

GIIGNL (International Group of Liquefied Gas Importers). 2014. World's LNG liquefaction plants and regasification terminals. p. 10. http://www.globallnginfo.com/World%20LNG%20Plants%20&%20Terminals.pdf (accessed February 20, 2015).

Ginsberg, S. 2013. Anti-fracturing mania reaches to the heart of Texas' Barnett Shale. *The American Oil & Gas Reporter*, October 2013:29.

Glasby, G.P. 2006. Abiogenic origin of hydrocarbons: An historical overview. Resource Geology 56:83–96.

GLNGI (Global LNG Info). 2014. World LNG trade 2013. http://www.globallnginfo.com/World%20LNG%20Trade%202013.pdf (accessed February 20, 2015).

Global Methane Initiative. 2014. Global methane emissions and mitigation opportunities. https://www.globalmethane.org/tools-resources/factsheets.aspx (accessed February 20, 2015).

Gobina, E. 2000. Gas-to-liquids Processes for Chemicals and Energy Production. Norwalk: Business Communications.

Gold, T. 1985. The origin of natural gas and petroleum, and the prognosis for future supplies. Annual Review of Energy 10:53–77.

Gold, T. 1993. The Origin of Methane (and Oil) in the Crust of the Earth. Washington, DC: USGS.

Gold, T. 1998. The Deep Hot Biosphere: Thy Myth of Fossil Fuels. Berlin: Springer.

Gold, R. 2014. The Boom: How Fracking Ignited the American Revolution and Changed the World. New York: Simon & Schuster.

Grace, J.D. and G.F. Hart. 1991. Urengoy gas field—U.S.S.R. West Siberian Basin, Tyumen District. In: Structural Traps III: Tectonic Fold and Fault Traps, Tulsa: AAPG, pp. 309–335.

Grandell, L., Hall, C.A.S. and M. Höök. 2011. Energy return on investment for Norwegian oil and gas from 1991 to 2008. Sustainability 3:2050–2070.

Green, M. 2014. Happy birthday, fracking! *Energy Tomorrow*, March 17, 2014. http://energytomorrow.org/blog/2014/march/march-17-happy-birthday-fracking (accessed February 20, 2015).

Groom, N. 2014. Chinese-backed Blu LNG slows down U.S. growth plans. *Reuters*, January 31, 2014. http://www.reuters.com/article/2014/01/31/us-blu-lng-retreat-idUSBREA0U1EZ20140131 (accessed February 20, 2015).

Guilford, M., Hall, C.A.S., Cleveland, C.J. 2011. New estimates of EROI for United States oil and gas, 1919–2010. Sustainability 3:1866–1887.

Guo, X. et al. 2014. Direct, nonoxidative conversion of methane to ethylene, aromatics, and hydrogen. Science 344:616–619.

Gurney, J. 1997. Migration or replenishment in the Gulf. *Petroleum Review* May 1997:200–203.

Gurt, M. 2014. Desert ceremony celebrates Turkmenistan-China gas axis. *Reuters*, May 7, 2014. http://www.reuters.com/article/2014/05/07/gas-turkmenistan-china-idUSL6N0NT2QS20140507 (accessed February 20, 2015).

Hackstein, J.H.P. and T.A. van Alen. 2010. Methanogens in the gastro-intestinal tract of animals. (Endo)symbiotic Methanogenic Archaea. Microbiology Monographs 19: 115–142.

Halbouty, M. 2001. Giant oil and gas fields of the decade 1990–2000. Paper Presented at AAPG Annual Convention, Denver, CO, June 3–6, 2001.

Hall, C.A.S. 2011. Introduction to special issue on new studies in EROI (Energy Return on Investment). Sustainability 3:1773–1777.

Halliburton. 2014. Fluids disclosure. http://www.halliburton.com/public/projects/pubsdata/Hydraulic_Fracturing/fluids_disclosure.html (accessed February 20, 2015).

Hammerschmidt, E.G. 1934. Formation of gas hydrates in natural gas transmission lines. Industrial Engineering & Chemistry 26:851.

Hand, E. 2014. Injection wells blamed in Oklahoma earthquakes. Science 345:13–14.

Harper, J.A. and J. Kostelnik. 2009. The Marcellus Shale Play in Pennsylvania. Harrisburg: Pennsylvania Geological Survey.

Harris, D., ed. 1996. Origins and Spread in Agriculture and Pastoralism in Eurasia. London: UCL Press.

Hatheway, A.W. 2012. Remediation of Former Manufactured Gas Plants and other Coal-Tar Sites. Boca Raton: CRC Press.

Hausmann, R. et al. 2013. The Atlas of Economic Complexity. Cambridge, MA: MIT Press.

Havens, J. 2003. Ready to Blow? Bulletin of the Atomic Scientists 59(4):16–18.

Hefley, W.E. et al. 2011. The Economic Impact of the Value Chain of a Marcellus Shale Well. Pittsburgh: University of Pittsburgh.

Hefner, R.A. III. 2009. The Grand Energy Transition. Hoboken: John Wiley & Sons, Inc..

Helm, D. 2014. A credible European security plan. Energy Futures Network Paper. http://www.dieterhelm.co.uk/sites/default/files/European%20Security%20Plan.pdf (accessed February 20, 2015).

Helman, C. 2012. Dow Chemical chief wants to limit U.S. LNG exports. http://www.forbes.com/sites/christopherhelman/2012/03/08/dow-chemical-chief-wants-to-limit-u-s-lng-exports/ (accessed February 20, 2015).

Hicklenton, P.R. 1988. $CO_2$ Enrichment in the Greenhouse: Principles and Practice. Portland: Timber Press.

Höglund-Isaksson, L. 2012. Global anthropogenic methane emissions 2005–2030: technical mitigation potentials and costs. Atmospheric Chemistry and Physics Discussions 12:11275–11315.

Högselius, P., Kaijser, A. and A. Åberg. 2010. Natural gas in cold war Europe: The making of a critical transnational infrastructure. http://www.palgraveconnect.com/pc/doifinder/view/10.1057/9781137358738.0007 (accessed March 11, 2015)

Hong, C.H. 2013. China's LNG trucks may rise fivefold by 2015. *Bloomberg*, March 14, 2013. http://www.bloomberg.com/news/2013-03-14/china-s-lng-trucks-may-increase-fivefold-by-2015-bernstein-says.html (accessed February 20, 2015).

Horncastle, A. et al. 2012. Future of Chemicals. Chicago: Booz & Company. http://www.geoexpro.com/articles/2010/04/the-king-of-giant-fields (accessed February 20, 2015).

Horton, S.T. 1995. Drill String and Drill Collars. Austin: Petroleum Extension Service, University of Texas at Austin.

Howard, G.C. 1949. Well Completion Process. US Patent 2,667,224, Washington, DC, A. Filing date June 29, 1949; publication date January 26, 1954. http://www.google.com/patents/US2667224 (accessed February 20, 2015).

Howarth, R.W., Santoro, R. and A. Ingraffea. 2011. Methane and the greenhouse-gas footprint of natural gas from shale formations. Climatic Change 106:679–690.

Howarth, R.W., Santoro, R. and A. Ingraffea. 2012. Venting and leaking of methane from shale gas development: response to Cathles et al. Climatic Change 113:537–549.

Hua, Y. et al. 2008. Sulige field in the Ordos Basin: Geological setting, field discovery and tight gas reservoirs. Marine and Petroleum Geology 25:387–400.

Hubbert, M.K. 1956. Nuclear energy and the fossil fuels. Presented at the Spring Meeting of the Southern Division of American Petroleum Institute, San Antonio, TX, March 7–9, 1956. http://www.hubbertpeak.com/hubbert/1956/1956.pdf (accessed February 20, 2015).

Hubbert, M.K. 1978. U.S. petroleum estimates, 1956–1978. Annual Meeting Papers, Division of Production, American Petroleum Institute, Denver, CO, April 2-5, 1978.

Hubert, C. and Q. Ragetly. 2013. GNVERT.GDF Sues promotes LNG as a fuel for heavy trucks in France by partnership with truck manufacturers. http://www.gastechnology.org/Training/Documents/LNG17-proceedings/7-3-Charlotte_Hubert.pdf (accessed February 20, 2015).

Hughes, S. 1871. A Treatise on Gas Works and the Practice of Manufacturing and Distributing Coal Gas. London: Lockwood & Company.

Hughes, J.D. 2013. A reality check on the shale evolution. Nature 494:307–308.

Hultman, N. et al. 2011. The greenhouse impact of unconventional gas for electricity generation. Environmental Research Letters 6 (October–December 2011):1–9.

Hunt, J.M. 1995. Petroleum Geochemistry and Geology. New York: W.H. Freeman & Company.

Hussain, Y. 2013. U.S. has 'overfracked and overdrilled,' Shell director says. *Financial Post*, October 18, 2013. http://business.financialpost.com/2013/10/18/u-s-has-overfracked-and-overdrilled-shell-director-says/?__lsa=8734-4302 (accessed February 20, 2015).

Hydrocarbons Technology. 2014. Sonatrach Skikda LNG Project. London: Hydrocarbons Technology.com.

IANGV (International Association of Natural Gas Vehicles). 2014. Current natural gas vehicle statistics. http://www.iangv.org/ (accessed February 20, 2015).

ICBA (International Carbon Black Association). 2014. Overview of uses. http://www.carbon-black.org/index.php/carbon-black-uses (accessed February 20, 2015).

ICF International. 2007. Energy trends in selected manufacturing sectors. http://www.epa.gov/sectors/pdf/energy/report.pdf (accessed February 20, 2015).

IEA (International Energy Agency). 2005. Unconventional oil & gas production. http://www.iea-etsap.org/web/E-TechDS/PDF/P02-Uncon%20oil&gas-GS-gct.pdf (accessed February 20, 2015).

IEA. 2007. Tracking industrial energy efficiency and $CO_2$ emissions. http://www.iea.org/publications/freepublications/publication/tracking_emissions.pdf (accessed February 20, 2015).

IEA. 2012. Golden Rules for a Golden Age of Gas. Paris: IEA. http://www.igu.org/news/igu-world-lng-report-2013.pdf (accessed February 20, 2015).

IEA. 2013. Key world energy statistics. http://www.iea.org/publications/freepublications/publication/tracking_emissions.pdf (accessed February 20, 2015).

IEA. 2014. World Energy Investment Outlook. Paris: IEA. http://www.iea.org/media/140603_WEOinvestment_Factsheets.pdf (accessed February 20, 2015).

IFDC (International Fertilizer Development Center). 2008. Worldwide Ammonia Capacity Listing by Plant. Muscle Shoals: IFDC.

IGU (International Gas Union). 2012. Geopolitics and natural gas. http://www.clingendaelenergy.com/inc/upload/files/Geopolitics_and_natural_gas_KL_final_report.pdf (accessed February 20, 2015).

IGU. 2014. World LNG Report—2014 Edition. Fornebu: IGU. http://www.igu.org/news/igu-world-lng-report-2013.pdf (accessed February 20, 2015).

IHRDC (International Human Resources Development Corporation). 2014. Gas processing and fractionation. http://www.ihrdc.com/els/po-demo/module14/mod_014_02.htm (accessed February 20, 2015).

IHS CERA. 2014. IHS CERA Upstream Capital Costs Index (UCCI). http://www.ihs.com/info/cera/ihsindexes/index.aspx (accessed February 20, 2015).

IMO (International Maritime Organization). 2008. International shipping and world trade facts and figures. http://www.imo.org/includes/blastData.asp/doc_id=8540/International%20Shipping%20and%20World%20Trade%20-%20facts%20and%20figures.pdf

Inman, M. 2014. Natural gas: The fracking fallacy. Nature 516:28–30. http://www.nature.com/news/natural-gas-the-fracking-fallacy-1.16430 (accessed March 10, 2015).

IPCC (Intergovernmental Panel on Climate Change). 2006. Stationary combustion In: 2006 IPCC Guidelines for National Greenhouse Gas Inventories. Geneva: Intergovernmental Panel on Climate Change. http://www.ipcc-nggip.iges.or.jp/public/2006gl/pdf/2_Volume2/V2_2_Ch2_Stationary_Combustion.pdf (accessed February 20, 2015).

IPCC. 2007. Climate Change 2007: Working Group I: The Physical Science Basis. Geneva: IPCC.

IPCC. 2013. Working group I contribution to the IPCC fifth assessment report climate change 2013: The physical science basis. http://www.climatechange2013.org/images/uploads/WGIAR5_WGI-12Doc2b_FinalDraft_Chapter08.pdf (accessed February 20, 2015).

IRGC (International Risk Governance Council). 2013. Risk Governance Guidelines for Unconventional Gas Development. Geneva: IRGC. http://www.irgc.org/wp-content/uploads/2013/12/IRGC-Report-Unconventional-Gas-Development-2013.pdf (accessed February 20, 2015).

Itar-Tass. 2014a. Russia's gas contract with China opens world market competition for Russian gas resources. http://en.itar-tass.com/economy/733974 (accessed February 20, 2015).

Itar-Tass. 2014b. China-bound pipeline to pump gas contracted by Gazprom and CNPC. Itar-Tass, June 27, 2014. http://en.itar-tass.com/economy/738083 (accessed February 20, 2015).

Jackson, R.B. et al. 2013. Increased stray gas abundance in a subset of drinking water wells near Marcellus shale gas extraction. Proceedings of the National Academy of Sciences 110:1125–11255.

Jaffe, A. and E. Morse. 2013. The end of OPEC. Foreign Policy, October 16, 2013. http://www.foreignpolicy.com/articles/2013/10/16/the_end_of_opec_america_energy_oil (accessed February 20, 2015).

Jarvis, D. et al. 1998. The association of respiratory symptoms and lung function with the use of gas for cooking. European Respiratory Journal 11:651–658.

Jiang, M. et al. 2011. Life Cycle Greenhouse Gas Emissions of Marcellus Shale Gas. Pittsburgh: Carnegie Mellon University and IOP Publishing. http://iopscience.iop. org/1748-9326/6/3/034014 (accessed February 20, 2015).

JOGMEC (Japan Oil, Gas and Metals Corporation). 2013. Gas production from methane hydrate layers confirmed. http://www.jogmec.go.jp/english/news/release/release0110. html (accessed February 20, 2015).

Jopson, B. 2014. US gas boom could be a geopolitical weapon. *Financial Times,* March 6, 2014. http://www.ft.com/cms/s/0/e2cf61ba-a489-11e3-b915-00144feab7de. html#axzz36nCxU4HK (accessed February 20, 2015).

Jordaan, S.M., Keith, D.W. and B. Stelfox. 2009. Quantifying land use of oil sands production: a life cycle perspective. *Environmental Research Letters* 4.024004. http:// iopscience.iop.org/1748-9326/4/2/024004/ (accessed February 20, 2015).

Kai, F.M. et al. 2011. Reduced methane growth rate explained by decreased Northern Hemisphere microbial sources. Nature 476:194–197.

Kaskey, J. 2013. Chemical companies rush to the U.S. thanks to cheap natural gas. *Bloomberg Businessweek,* July 25, 2014. http://www.businessweek.com/articles/ 2013-07-25/chemical-companies-rush-to-the-u-dot-s-dot-thanks-to-cheap-natural-gas (accessed February 20, 2015).

Kaufman, S. 2013. Project Plowshare: The Peaceful Use of Nuclear Explosives in Cold War America. Ithaca: Cornell University Press.

Kaur, S. 2010. Asian and European bunker and residual markets. http://www.platts.com/ im.platts.content/productsservices/conferenceandevents/2013/pc322/presentations/ sharmilpal_kaur_day2.pdf (accessed February 20, 2015).

KBR (Kellogg Brown & Root). 2014. Kellogg Brown & Root. http://www.kbr.com/ (accessed February 20, 2015).

Kehlhofer, R., Rukes, B. and F. Hannemann. 2009. Combined-Cycle Gas & Steam Turbine Power Plants. Tulsa: Pennwell Books.

Kenney, J.F. et al. 2002. The evolution of multicomponent systems at high pressures: VI. The thermodynamic stability of the hydrogen–carbon system: The genesis of hydrocarbons and the origin of petroleum. Proceedings of the National Academy of Sciences 99:10976–10981.

Kenny, J.F. et al. 2009. Estimated use of water in the United States in 2005. http://pubs. usgs.gov/circ/1344/ (accessed February 20, 2015).

Kharaka, Y.K. et al. 2013. The energy-water nexus: potential groundwater-quality degradation associated with production of shale gas. Procedia Earth and Planetary Science 7:417–422.

Kiefner, J.F. and M.J. Rosenfeld. 2012. The Role of Pipeline Age in Pipeline Safety. Worthington: The INGAA Foundation. http://www.ingaa.org/File.aspx?id=19307 (accessed February 20, 2015).

Kiefner, J.F. and C.J. Trench. 2001. Oil pipeline characteristics and risk factors: Illustrations from the decade of construction. http://www.api.org/oil-and-natural-gas-overview/ transporting-oil-and-natural-gas/pipeline-performance-ppts/ppts-related-files/~/media/ files/oil-and-natural-gas/ppts/other-files/decadefinal.ashx (accessed February 20, 2015).

King, H. 2014. Production and Royalty Declines in a Natural Gas Well Over Time. Geology.com. http://geology.com/royalty/production-decline.shtml (accessed March 10, 2015).

Kintisch, E. 2014. Hunting a climate fugitive. Science 344:1472–1473.

Kirschke, S. et al. 2013. Three decades of global methane sources and sinks. Nature Geoscience 6:813–823.

Klett, T.R. et al. 2011. Assessment of undiscovered oil and gas resources of the Amu Darya Basin and Afghan–Tajik Basin Provinces, Afghanistan, Iran, Tajikistan, Turkmenistan, and Uzbekistan, 2011. http://pubs.usgs.gov/fs/2011/3154/ (accessed February 20, 2015).

Klett, T.R. et al. 2012. Assessment of potential additions to conventional oil and gas resources of the world (outside the United States) from reserve growth, 2012. http://pubs.usgs.gov/fs/2012/3052/fs2012-3052.pdf (accessed February 20, 2015).

"K" Line. 2014. Development of a compressed natural gas carrier. https://www.kline.co.jp/en/service/energy/cng/detail.html (accessed February 20, 2015).

Koch, H. 1965. 75 Jahre Mannesmann Geschichte einer Erfindung und eines Unternehmens, 1890–1965. Düsseldorf: Mannesmann AG.

Komlev, S. 2014. EU-Russian Gas Trade: Too Deep and Comprehensive to Fail. Moscow: Moscow Energy Charter Forum. http://www.encharter.org/fileadmin/user_upload/Conferences/2014_April_3/1-2_Komlev.pdf (accessed February 20, 2015).

Kudryavtsev, N.A. 1959. Oil, Gas, and Solid Bitumens in Igneous and Metamorphic Rocks. Leningrad: State Fuel Technical Press.

Kumar, S. et al. 2011. Current status and future projections of LNG demand and supplies: A global perspective. Energy Policy 39:4097–4104.

Kuuskraa, V.A. 2004. Natural gas resources, unconventional. In: Encyclopedia of Energy, C. Cleveland et al. eds., San Diego: Elsevier, Vol. 4, pp. 257–272.

Kuuskraa, V.A. 2013. EIA/ARI world shale gas and shale oil resource assessment. http://www.eia.gov/conference/2013/pdf/presentations/kuuskraa.pdf (accessed February 20, 2015).

Kuuskraa, V.A. et al. 2011. World Shale Gas Resources: An Initial Assessment of 14 Regions Outside the United States. Arlington: Advanced Resources International. http://www.adv-res.com/pdf/ARI%20EIA%20Intl%20Gas%20Shale%20APR%202011.pdf (accessed February 20, 2015).

Kvenvolden, K.A., 1988. Methane hydrate—A major reservoir of carbon in the shallow geosphere? Chemical Geology 71:41–51.

Kvenvolden, K.A., 1993. Gas hydrates—Geological perspective and global change. Review of Geophysics 31:173l–187l.

Laherrère, J.H. 2000. Global natural gas perspectives. http://www.hubbertpeak.com/laherrere/ngperspective/ (accessed February 20, 2015).

Lamlom, S. H., R. A. Savidge. 2003. A reassessment of carbon content in wood: Variation within and between 41 North American species. Biomass and Bioenergy 25:381–388.

Langston, L.S. 2013. The adaptable gas turbine. American Scientist 101:264–267.

Laubach, S.E. et al. 1998. Characteristics and origins of coal cleat: A review. International Journal of Coal Geology 35:175–207.

Lechtenböhmer, S. et al. 2007. Tapping the leakages: Methane losses, mitigation options and policy issues for Russian long distance gas transmission pipelines. International Journal of Greenhouse Gas Control 1:387–395.

Levi, M. 2013. The Power Surge: Energy, Opportunity and Battle for America's Future. New York: Oxford University Press.

Lewis, J.A. 1961. History of Petroleum Engineering. New York: API.

Li, G. 2011. World Atlas of Oil and Gas Basins. Chichester: Wiley-Blackwell.

Li, M. 2014. Geophysical Exploration Technology: Applications in Lithological and Stratigraphic Reservoirs. Amsterdam: Elsevier.

Linde, C.P. 1916. Aus meinem Leben und von meiner Arbeit. München: R. Oldenbourg.

Linde. 2013. LNG technology. http://www.linde-engineering.com/internet.global. lindeengineering.global/en/images/LNG_1_1_e_13_150dpi19_4577.pdf (accessed February 20, 2015).

Linde. 2014. LNG in transportation. http://lindelng.com/index.php/products-and-services/lng-in-transportation (accessed February 20, 2015).

LNG Canada. 2014. Why did LNG Canada choose to locate its project in Katimat? http://lngcanada.ca/faq-items/why-did-lng-canada-choose-to-locate-its-project-in-kitimat/ (accessed February 20, 2015).

Loder, A. 2014a. Shale gas drillers feast on junk debt to stay on treadmill. http://www.bloomberg.com/news/2014-04-30/shale-drillers-feast-on-junk-debt-to-say-on-treadmill.html (accessed February 20, 2015).

Loder, A. 2014b. Shakeout threatens shale patch as frackers go for broke. http://www.bloomberg.com/news/2014-05-26/shakeout-threatens-shale-patch-as-frackers-go-for-broke.html (accessed February 20, 2015).

Logue, J.M. et al. 2013. Pollutant exposures from natural gas cooking burners: A simulation-based assessment for Southern California. Environmental Health Perspectives 122:43–50.

Lollar, B.S. et al. 2002. Abiogenic formation of alkanes in the Earth's crust as a minor source for global hydrocarbon reservoirs. Nature 416:522–524.

Loree, D., Byrd, A. and R. Lestz. 2014. WaterLess fracturing technology. http://www.siltnet.net/documents/lestz_condensed_version_2_15_11.pdf (accessed February 20, 2015).

Lovelock, J.E. and L. Margulis. 1974. Atmospheric homeostasis by and for the biosphere: The gaia hypothesis. Tellus 26:1–10.

Lowrie, A. and M.D. Max. 1999. The extraordinary promise and challenge of gas hydrates. World Oil 49(50):53–55.

Magoon, L.B. and W.G. Dow, eds. 1994. The Petroleum System—From Source to Trap. AAPG Memoir 60. Washington, DC: USGS.

Magoon, L.B. and J.W. Schmoker. 2000. The Total Petroleum System—The Natural Fluid Network that Constrains the Assessment Unit. Washington, DC: USGS.

Mahfoud, R. F. and J. N. Beck. 1995. Why the Middle East fields may produce oil forever. *Offshore* April 1995:58–64, 106.

Makogon, Y. F. 1965. A gas hydrate formation in the gas saturated layers under low temperature. Gas Industry 5:14–15.

Makogon, Y.F., Holditch, S.A., and T.Y. Makogon. 2007. Natural gas-hydrates—A potential energy source for the 21st Century. Journal of Petroleum Science and Engineering 56:14–31.

Mallinson, R.G. 2004. Natural gas processing and products. In: Encyclopedia of Energy, C. Cleveland et al. eds., San Diego: Elsevier, Vol. 4, pp. 235–247.

Mankin, C.J. 1983. Unconventional sources of natural gas. Annual Review of Energy 8:27–43.

Mann, P., Gahagan, L. and M.B. Gordon. 2001. Tectonic setting of the world's giant oil fields. World Oil, 222(9). http://www.worldoil.com/September-2001-Tectonic-setting-of-the-worlds-giant-oil-fields.html (accessed February 20, 2015).

Marbek. 2010. Study of opportunities for natural gas in the transportation sector. http://www.cngva.org/media/4302/marbek_ngv_final_report-april_2010.pdf (accessed February 20, 2015).

Marchetti, C. 1977. Primary energy substitution models: In the interaction between energy and society. Technological Forecasting and Social Change 10:345–356.

Marchetti, C. and N. Nakićenović. 1979. The Dynamics of Energy Systems and the Logistic Substitution Model. Vienna: IIASA.

Marine Exchange of Alaska. 2014. Shipping cook inlet liquefied natural gas—The tankers. http://www.mxak.org/home/photo_essays/lng_tankers.html; http://www.hydrocarbons-technology.com/projects/sonatrach/ (accessed February 20, 2015).

Marine Traffic. 2014. Mozah. http://www.marinetraffic.com/ais/details/ships/9337755/vessel:MOZAH (accessed February 20, 2015).

Marsters, P.V. 2013. A revolution on the horizon. China Environment Series 2012/2013:35–47.

Martin, S. 2010. Extra ethane poses obstacle in US Marcellus Shale development. http://www.icis.com/resources/news/2010/10/18/9402406/extra-ethane-poses-obstacle-in-us-marcellus-shale-development/ (accessed February 20, 2015).

Matthews & Associates. 2014. $3 million verdict in Texas fracking case. http://www.dmlawfirm.com/3-million-verdict-fracking-case (accessed February 20, 2015).

Maugeri, L. 2013. Beyond the Age of Oil. Santa Barbara: Praeger.

McCarthy, K. et al. 2011. Basic petroleum chemistry for source rock evaluation. Oilfield Review Summer 2011:32–43.

McJeon, H. et al. 2014. Limited impact of decadal-scale climate change from increased use of natural gas. Nature 514:482–485.

McKelvey, V.E. 1973. Mineral resource estimates and public policy. In: United States Mineral Resources, D.A. Brobst and W.P. Pratt, eds., Washington, DC: USGS, pp. 9–19.

McKibben, B. et al. 2014. A letter to President Obama to stop the disastrous rush to export fracked gas at Cove Point and nationwide. http://org.salsalabs.com/o/423/images/LNG-Export-PresidentObama-Climate-Letter31814.pdf (accessed February 20, 2015).

McNutt, M. 2013. Bridge or crutch? Science 342:909.

Messner, J. and G. Babies. 2012. Transport of natural gas. http://www.shell.com/global/future-energy/scenarios/new-lens-scenarios.html (accessed February 20, 2015).

Methanex. 2014. Methanex investor presentation. http://freepdfs.net/methanex-investor-presentation/99fb9fe1efad0b415408f088faa45a8a/ (accessed March 11, 2015).

Milkov, A.V. 2010. Methanogenic biodegradation of petroleum in the West Siberian Basin (Russia): Significance for formation of giant Cenomanian gas pools. AAPG Bulletin 94:1485–1541.

MIT (Massachusetts Institute of Technology). 2011. The Future of Natural Gas: An Interdisciplinary MIT Study. Cambridge, MA: MIT. http://web.mit.edu/ceepr/www/publications/Natural_Gas_Study.pdf (accessed February 20, 2015).

Molina, J. et al. 2011. Molecular evidence for a single evolutionary origin of domesticated rice. Proceedings of the National Academy of Sciences 108:8351–8356.

Montgomery, C.T. and M.B. Smith. 2010. Hydraulic fracturing: History of an enduring technology. Journal of Petroleum Technology December 2010:26–32. http://www.ourenergypolicy.org/wp-content/uploads/2013/07/Hydraulic.pdf (accessed February 20, 2015).

Murphy, S.L., Xu, J. and K.D. Kochanek. 2013. Deaths: Final data for 2010. National Vital Statistics Reports 61(4):1–117.

NAM (Nederlandse Aardolie Maatschappij). 2009. Groningen gas field. http://www-static.shell.com/content/dam/shell/static/nam-en/downloads/pdf/flyer-namg50eng.pdf (accessed February 20, 2015).

NaturalGas.org. 2014. The history of regulation. http://naturalgas.org/regulation/history/ (accessed February 20, 2015).

Navarro, M. 2013. Bans and rules muddy prospects for gas drilling. *New York Times*, January 3, 2013:A18.

NEB (National Energy Board). 2013. Montney formation one of the largest gas resources in the world, report shows. http://www.neb-one.gc.ca/clf-nsi/rthnb/nws/nwsrls/2013/nwsrls30-eng.html (accessed February 20, 2015).

NEB. 2014a. Canada's oil sands. http://www.neb.gc.ca/clf-nsi/rnrgynfmtn/nrgyrprt/lsnd/pprtntsndchllngs20152006/qapprtntsndchllngs20152006-eng.html (accessed February 20, 2015).

NEB. 2014b. Canadian Pipeline Transportation System. Calgary: NEB. http://www.neb-one.gc.ca/clf-nsi/rnrgynfmtn/nrgyrprt/trnsprttn/2014trnsprttnssssmnt/2014trnsprttnss ssmnt-eng.pdf (accessed February 20, 2015).

Needham, J. 1964. The Development of Iron and Steel Technology in China. Cambridge: W. Heffer & Sons.

NETL (National Energy Technology Laboratory). 2014. Fire in the ice. http://www.netl.doe.gov/research/oil-and-gas/methane-hydrates/fire-in-the-ice (accessed February 20, 2015).

NETL. 2013. Modern shale gas development in the United States: An update. http://www.netl.doe.gov/File%20Library/Research/Oil-Gas/shale-gas-primer-update-2013.pdf (accessed February 20, 2015).

NGMA (National Greenhouse Manufacturers Association). 2014. Carbon dioxide enrichment. http://www.shell.com/global/future-energy/scenarios/new-lens-scenarios.html (accessed February 20, 2015).

Nilsen, Ø. 2012. Snøhvit introduction to Melkøya plant. http://www02.abb.com/global/abbzh/abbzh250.nsf/0/56a5a9fc590db243c1257a2100342427/$file/Press_trip_Hammerfest_Presentation_Introduction+to+Melk%C3%B8ya+plant+-+%C3%98 ivind+Nilsen.pdf (accessed February 20, 2015).

Nisbet, E.G., Dlugokencky, E.J. and P. Bousquet. 2014. Methane on the rise—Again. Science 343:493–494.

NL Oil and Gas Portal. 2014. Groningen gas field. http://www.nlog.nl/en/reserves/Groningen.html (accessed February 20, 2015).

Noble, A.C. 1972. The Wagon wheel project. http://www.wyohistory.org/essays/wagon-wheel-project (accessed February 20, 2015).

Nord Stream. 2014. Nord stream: The new gas supply route for Europe. http://www.nord-stream.com/ (accessed February 20, 2015).

Novatek. 2014a. South-Tambeyskoye Field (Yamal LNG Project). http://www.novatek.ru/en/business/yamal/southtambey/ (accessed February 20, 2015).

Novatek. 2014b. Classification of reserves. http://www.novatek.ru/en/press/reserves/ (accessed February 20, 2015).

NPC (National Petroleum Council). 2011. Prudent Development: Realizing the Potential of North America's Abundant Natural Gas and Oil Resources. Washington, DC: NPC. http://www.ourenergypolicy.org/wp-content/uploads/2013/07/Hydraulic.pdf (accessed February 20, 2015).

NYSDEC (New York State Department of Environmental Conservation). 2009. Revised draft SGEIS on the oil, gas and solution mining regulatory program (September 2011). http://www.dec.ny.gov/energy/75370.html (accessed February 20, 2015).

Ohnsman, A. 2013. Hydrogen prototype takes to the road in race toward fuel cells. http://www.japantimes.co.jp/news/2013/10/11/business/corporate-business/hydrogen-prototype-takes-to-the-road-in-race-toward-fuel-cells/#.U9FTwbkg-M8 (accessed February 20, 2015).

Olah, G.A., Goeppert, A. and G.K.S. Prakash. 2006. Beyond Oil and Gas: The Methanol Economy. Weinheim: Wiley-VCH.

Ortwein, S.N. 2013. 50 years of 3D seismic, and more to come. In: World Petroleum Council, 80th Anniversary Edition, pp. 70–73. London: First World Petroleum. http://www.world-petroleum.org/docs/docs/publications/wpc%2080th%201.pdf (accessed February 20, 2015).

Osgouei, R.E. and M. Sorgun. 2012 A critical evaluation of Iranian natural gas resources. Energy Sources Part B 7:113–120.

O'Sullivan, F. and S. Paltsev. 2012. Shale Gas Production: Potential versus Actual Greenhouse Gas Emissions. Cambridge, MA: MIT. http://globalchange.mit.edu/files/document/MITJPSPGC_Rpt234.pdf (accessed February 20, 2015).

Palkovits, R. et al. 2009. Solid catalysts for the selective low-temperature oxidation of methane to methanol. Angewandte Chemie International Edition 48:6909–6912.

Patzek, T.W., Male, F. and M. Marder. 2013. Gas production in the Barnett Shale obeys a simple scaling theory. PNAS 110:19731–19736.

Peck, B. and H. van der Velde. 2013. A High Capacity Floating LNG Design. Houston: LNG17, April 19, 2013. http://www.gastechnology.org/Training/Documents/LNG17-proceedings/12_01-Harry-vanderVelde-Shell-Presentation.pdf (accessed February 20, 2015).

Periana, R.A. et al. 1998. Platinum catalysts for the high-yield oxidation of methane to a methanol derivative. Science 280:560–564.

Peters, K., Schenk, O. and B. Wygrala, 2009. Exploration paradigm shift: The dynamic petroleum system concept. Swiss Bulletin für angewandte Geologie 14:56–71. http://www.angewandte-geologie.ch/Dokumente/Archiv/Vol14_1_2/1412_5Petersetal.pdf (accessed February 20, 2015).

Pétron, G., et al. 2012. Hydrocarbon emissions characterization in the Colorado Front Range: A pilot study. Journal of Geophysical Research-Atmospheres 117(D4):D04304.

Pew Research Center. 2013. Energy: Key data points. http://www.pewresearch.org/key-data-points/energy-key-data-points/ (accessed February 20, 2015).

PHMSA (Pipeline & Hazardous Materials Safety Administration). 2014a. Annual report mileage for natural gas transmission & gathering systems. http://www.phmsa.dot.gov/portal/site/PHMSA/menuitem.6f23687cf7b00b0f22e4c6962d9c8789/?vgnextoid=78e4f5448a359310VgnVCM1000001ecb7898RCRD&vgnextchannel=3430fb649a2dc110VgnVCM1000009ed07898RCRD&vgnextfmt=print&vgnextnoice=1 (accessed February 20, 2015).

PHMSA. 2014b. Significant pipeline incidents. http://primis.phmsa.dot.gov/comm/reports/safety/sigpsi.html (accessed February 20, 2015).

Pickens, T.B. 2008. The plan. http://www.pickensplan.com/the-plan/ (accessed February 20, 2015).

Platts. 2014. UDI Combined-Cycle and Gas Turbine Data Set (CCGT), 2013 Edition. New York: Platts.

PNG LNG. 2014. PNG LNG. http://www.pnglng.com/ (accessed February 20, 2015).

Porfir'yev, V.B. 1959. The Problem of Migration of Petroleum and the Formation of Accumulations of Oil and Gas. Moscow: Gostoptekhizdat.

Porfir'yev, V.B. 1974. Inorganic origin of petroleum. AAPG Bulletin 58:3–33.

Potential Gas Committee. 2013. The Report of the Potential Gas Committee. Golden: Potential Gas Committee. http://potentialgas.org/ (accessed February 20, 2015).

Powell, B. 2013. Liners provide secondary containment. *The American Oil & Gas Reporter* December 2012:87–92.

PW Power Systems. 2014. PW power systems. http://www.pwps.com/ (accessed February 20, 2015).

PWC (Price Waterhouse Cooper). 2012. Shale gas reshaping the US chemical industry. http://www.pwc.com/en_US/us/industrial-products/publications/assets/pwc-shale-gas-chemicals-industry-potential.pdf (accessed February 20, 2015).

Qatargas. 2014a. Qatargas fleet. https://www.qatargas.com/English/AboutUs/Shipping/Pages/default.aspx (accessed February 20, 2015).

Qatargas. 2014b. Current operations. http://www.qatargas.com/English/AboutUs/Pages/CurrentOperationsD.aspx (accessed February 20, 2015).

Quik Water. 2014. The world's cleanest, greenest direct-contact water heating system is still the original. http://www.gmitchell.ca/Quikwater/Product%20Litterature/QuikWater%20-%20Brochure%20-%20Flagship.PDF (accessed February 20, 2015).

Rao, A. 2012. Combined Cycle Systems for Near-Zero Emission Power Generation. Cambridge: Woodhead Publishing.

Rascoe, A. and E. O'Grady. 2009. Pickens delays wind farm on finance, grid issues. *Reuters*, July 8, 2009 http://www.reuters.com (accessed March 10, 2015).

Rasmussen B. 2000. Filamentous microfossils in a 3,235-million-year-old volcanogenic massive sulphide deposit. Nature 405:676–679.

Ratner, M. et al. 2013. U.S. Natural Gas Exports: New Opportunities, Uncertain Outcomes. Washington, DC: US Library of Congress.

Reig, P., Luo, T. and J.N. Proctor. 2014. Global Shale Gas Development: Water Availability & Business Risks. Washington, DC: World Resources Institute.

Reith, F. 2011. Life in the deep subsurface. Geology 39:287–288.

Reshetnikov, A.I., Paramonova N.N. and A.A. Shashkov. 2000. An evaluation of historical methane emissions from the Soviet gas industry. Journal of Geophysical Research 105:3517–3529.

Reuters. 2014. UPDATE 2-U.S. EIA cuts recoverable Monterey shale oil estimate by 96 pct. http://www.reuters.com/article/2014/05/21/eia-monterey-shale-idUSL1N0O713N20140521 (accessed February 20, 2015).

Rice, D.D. 1997. Coalbed methane—An untapped energy resource and an environmental concern. U.S. Geological Survey Fact Sheet FS–019–97.

Roland, T.H. (2010). Associated Petroleum Gas in Russia: Reasons for non-utilization. FNI Report 13/2010. http://www.fni.no/doc&pdf/FNI-R1310.pdf (accessed March 9, 2015).

Rotty, R.M. 1974. First estimates of global flaring of natural gas. Atmospheric Environment 8:681–686.

Royal Dutch Shell. 2013. New lens scenarios. http://www.shell.com/global/future-energy/scenarios/new-lens-scenarios.html (accessed February 20, 2015).

Ruddiman, W.F. 2005. Plows, Plagues & Petroleum. Princeton: Princeton University Press.

Sabine Pipe Line LLC. 2014. The Henry hub. http://www.sabinepipeline.com/Home/Report/tabid/241/default.aspx?ID=52 (accessed February 20, 2015).

Salzgitter Mannesmann. 2014. Salzgitter Mannesmann International. http://www.salzgitter-mannesmann-international.com/ (accessed February 20, 2015).

Sandrea, R. 2012. Evaluating production potential of mature US oil, gas shale plays. *Oil & Gas Journal*. http://www.ogj.com/articles/print/vol-110/issue-12/exploration-development/evaluating-production-potential-of-mature-us-oil.html (accessed February 20, 2015).

Sandrea, R. 2014. US Shale Gas and Tight Oil Industry Performance: Challenges and Opportunities. Oxford: The Oxford Institute for Energy Studies. http://www.oxfordenergy.org/wpcms/wp-content/uploads/2014/03/US-shale-gas-and-tight-oil-industry-performance-challenges-and-opportunities.pdf (accessed February 20, 2015).

Sasol. 2014. Sasol. http://www.sasol.com/ (accessed February 20, 2015).

Scanlon, B.R., Duncan, I. and R.C. Reedy. 2013. Drought and the water–energy nexus in Texas. Environmental Research Letters 8(2013):045033.

Schenk, C.J. 2012. An estimate of undiscovered conventional oil and gas resources of the world, 2012. http://pubs.usgs.gov/fs/2012/3042/fs2012-3042.pdf (accessed February 20, 2015).

Schenk, C.J. and R.M. Pollastro. 2002. Natural Gas Production in the United States. Reston: USGS.

Schmoker, J.W. Krystinik, K.B. and R.B. Halley. 1985. Selected characteristics of limestone and dolomite reservoirs in the United States. AAPG Bulletin 69:733–741.

Schurr, S. H. and B. C. Netschert. 1960. Energy in the American Economy 1850–1975. Baltimore: Johns Hopkins University Press.

Schwerin, B.1991. Die Treibstoffversorgung durch Kohlehydrierung in Deutschland von 1933 bis 1945, unter besonderer Berücksichtigung wirtschafts- und energiepolitischer Einflüsse. Köln: Müller Botermann.

Seele, R. 2012. On the path to a greener future. http://www.wintershall.no/news-media/press-releases/detail/2012/3/29/on-the-path-to-a-greener-future.html (accessed February 20, 2015).

Selley, R.C. 1997. Elements of Petroleum Geology. San Diego: Academic Press.

Semolinos, P. 2013. LNG as bunker fuel: Challenges to be overcome. http://www.wintershall.no/news-media/press-releases/detail/2012/3/29/on-the-path-to-a-greener-future.html (accessed February 20, 2015).

Sephton, M.A. and R.M. Hazen. 2013. On the origins of deep hydrocarbons. Reviews in Mineralogy & Geochemistry 75:449–465.

Service, R. 2014. The bond breaker. Science 344:1474–1475.

Shale Gas International. 2014. China cuts down its shale estimates while investing in coal gasification. http://www.shalegas.international/2014/08/07/china-cuts-down-its-shale-estimates-while-investing-in-coal-gassification/ (accessed February 20, 2015).

Shell. 2013. Shell to charter barges powered solely by LNG. http://www.shell.com/global/products-services/solutions-for-businesses/shipping-trading/about-shell-shipping/lng-barges-05092012.html (accessed February 20, 2015).

Shell. 2014. Pearl GTL—An overview. http://www.shell.com/global/aboutshell/major-projects-2/pearl/overview.html (accessed February 20, 2015).

Siemens. 2014. Siemens gas turbines. http://www.energy.siemens.com/hq/en/fossil-power-generation/gas-turbines/ (accessed February 20, 2015).

Simakov, S.N. 1986. Forecasting and Estimation of the Petroleum-Bearing Subsurface at Great Depths. Leningrad: Nedra.

Sjaastad, A.K., Jørgensen, R.B. and K. Svendsen. 2010. Exposure to polycyclic aromatic hydrocarbons (PAHs), mutagenic aldehydes and particulate matter during pan frying of beefsteak. Occupational and Environmental Medicine 67:228–232.

Skarke, A. et al. 2014. Widespread methane leakage from the sea floor on the northern US Atlantic margin. Nature Geoscience 7:657–661.

Skone, T.J. 2011. Life cycle greenhouse gas analysis of natural gas extraction, delivery and electricity production in the United States. http://www.capp.ca/getdoc.aspx? DocId=215278 (accessed February 20, 2015).

Slatt, R.M. 2006. Stratigraphic reservoir characterization for petroleum geologists, geophysicists, and engineers. In: Handbook of Petroleum Exploration and Production. Amsterdam: Elsevier, Vol. 6, pp. 275–305.

Slattery, D. and B. Hutchinson. 2014. NTSB, known for probing air crashes, is investigating Harlem's deadly gas-pipe explosion. New York Daily News, March 16, 2014. http://www.nydailynews.com/new-york/uptown/air-crash-agency-investigates-harlem-pipe-explosion-article-1.1719970#ixzz2vqoInJuy (accessed February 20, 2015).

Smeenk. T. 2010. Russian Gas for Europe: Creating Access and Choice. The Hague: Clingendael International Energy Programme.

Smil, V. 2000. Feeding the World. Cambridge, MA: MIT Press.

Smil, V. 2001. Enriching the Earth: Fritz Haber, Carl Bosch, and the Transformation of World Food Production. Cambridge, MA: MIT Press.

Smil, V. 2003. Energy at the Crossroads. Cambridge, MA: MIT Press.

Smil, V. 2004. China's Past China's Future. London: RoutledgeCurzon.

Smil, V. 2006. Transforming the Twentieth Century. New York: Oxford University Press.

Smil, V. 2008. Oil: A Beginner's Guide. Oxford: Oneworld.

Smil, V. 2010a. Prime Movers of Globalization: The History and Impact of Diesel Engines and Gas Turbines. Cambridge, MA: MIT Press.

Smil, V. 2010b. Energy Transitions; History, Requirements, Prospects. Santa Barbara: Praeger.

Smil, V. 2013a. Making the Modern World. Oxford: Wiley.

Smil, V. 2013b. Made in the USA: The Rise and Retreat of American Manufacturing. Cambridge, MA: MIT Press.

Smil, V. 2014. The long slow rise of solar and wind. Scientific American 282 (1):52–57.

Smil, V. 2015. Power Density: A Key to Understanding Energy Sources and Uses. Cambridge, MA: MIT Press.

Song, Y. 1673. Tiangong kaiwu (The Creations of Nature and Man). Translated by Sun, E. and S. Sun. Pennsylvania State University Press, State College, PA (1966).

Songhurst, B. 2014. LNG Plant Cost Escalation. Oxford: The Oxford Institute for Energy Studies. http://www.oxfordenergy.org/wpcms/wp-content/uploads/2014/02/NG-83.pdf (accessed February 20, 2015).

Sorenson, R.P. 2005. A dynamic model for the Permian Panhandle and Hugoton fields, western Anadarko basin. AAPG Bulletin 89:921–938.

Sorkhabi, R. 2010. The king of giant fields. GeoExPro, 7(4). http://www.geoexpro.com/articles/2010/04/the-king-of-giant-fields (accessed February 20, 2015).

South Stream. 2014. South stream. http://www.gazprom.com/about/production/projects/pipelines/south-stream/ (accessed March 11, 2015).

Speight, J.G. 2007. Natural Gas: A Basic Handbook. Houston: Gulf Publishing.

Speight, J.G. 2013. Shale Gas Production Processes. Houston: Gulf Professional Publishing.

Spiegel. 2013. Full throttle ahead: US tips global power scales with fracking. Der Spiegel, February 2, 2013. http://www.spiegel.de/international/world/new-gas-extraction-methods-alter-global-balance-of-power-a-880546.html (accessed February 20, 2015).

Stephenson, T., Valle, J.E. and Riera-Palou, X. 2011. Modeling the relative GHG emissions of conventional and shale gas production. Environmental Science & Technology 45:10757–10764.

Stern, D.I. and R.K. Kaufmann. 2001. Annual Estimates of Global Anthropogenic Methane E: 1860-1994. Oak Ridge: ORNL.

Steward, D.B., 2007. The Barnett Shale Play: Phoenix of the Fort Worth Basin: A History. Fort Worth: Fort Worth and North Texas Geological Societies.

Stokstad, E. 2014. Will fracking put too much fizz in your water? Science 344:1468–1471.

Stolper, D.A. et al. 2014. Formation temperatures of thermogenic and biogenic methane. Science 344:1500–1503.

Subsea Oil and Gas Directory. 2014. Subsea oil gas pipelines in the North Sea Area. http://www.subsea.org/pipelines/allbyarea.asp (accessed February 20, 2015).

Taylor, C. 2002. Formation studies of methane hydrates with surfactants. 2nd International Workshop on Methane Hydrates, Washington, DC, October 2002.

TEPCO (Tokyo Electric Power Company). 2013. Thermal power generation. http://www.tepco.co.jp/en/challenge/energy/thermal/power-g-e.html (accessed February 20, 2015).

TETCO (Texas Eastern Transmission Corporation). 2000. The Big Inch and Little Big Inch Pipelines. Houston: TETCO. http://historicmonroe.org/labor/pix/inchlines.pdf (accessed February 20, 2015).

TNO (Toegepast Natuurwetenschappelijk Onderzoek). 2007. Estimation of emissions of fine particulate matter ($PM_{2.5}$) in Europe. http://ec.europa.eu/environment/air/pollutants/pdf/report_2007_ar0322.pdf

TNO. 2008. Greenhouse horticulture. https://www.google.ca/#q=tno+greenhouse+horticulture (accessed February 20, 2015).

Tobin, J. 2003. Natural gas market centers and hubs—2003 update. http://www.eia.gov/pub/oil_gas/natural_gas/feature_articles/2003/market_hubs/mkthubsweb.html (accessed February 20, 2015).

Tollefson, J. 2012. Air sampling reveals high emissions from gas field. Nature 482:139–140.

Tollefson, J. 2013. China slow to tap shale-gas bonanza. Nature 494:294.

Tollefson, J. 2014. Climate change: The case of the missing heat. Nature 505:276–278.

Total. 2007. Tight gas reservoirs: Technology-intensive resources. http://www.total.com/sites/default/files/atoms/file/total-tight-gas-reservoirs-technology-intensive-resources (accessed February 20, 2015).

Total. 2014. St. Fergus gas terminal. http://www.uk.total.com/activities/st_fergus_terminal.asp (accessed February 20, 2015).

Trading Economics. 2014. Japan balance of trade. http://www.tradingeconomics.com/japan/balance-of-trade (accessed February 20, 2015).

TransCanada. 2014. Wholly owned pipelines. http://www.transcanada.com/natural-gas-pipelines.html#CM (accessed February 20, 2015).

Truck News. 2014. News. http://www.trucknews.com/news/ (accessed February 20, 2015).

Ulmishek, G.F. 2004. Petroleum Geology and Resources of the Amu-Darya Basin, Turkmenistan, Uzbekistan, Afghanistan, and Iran. Reston: USGS.

UNDESA (United Nation Department of Economic and Social Affairs). 2013. Detailed indicators. http://unstats.un.org/unsd/databases.htm

Uniongas. 2013. Natural gas cooling solutions: An overview of the technology and opportunities. http://energy.gov/sites/prod/files/2013/03/f0/ShaleGasPrimer_Online_4-2009.pdf (accessed February 20, 2015).

United State Congress. 1973. Nuclear Stimulation of Natural Gas. Hearing, Ninety-third Congress, First Session, May 11, 1973. Washington, DC: USGPO.

UNO (United Nations Organization). 1976. World Energy Supplies 1950-1974. New York: UNO.

Upham, C.W., ed. 1851. The life of General Washington: First President of the United States. Volume 2. London: National Illustrated Library.

USBM (US Bureau of Mines). 1946. Report on the Investigation of the fire at the liquefaction, storage, and regasification plant of East Ohio Gas Co., Cleveland, OH, October 20, 1944. Washington, DC: USBM.

USCB (US Census Bureau). 1975. Historical Statistics of the United States: Colonial Times to 1970. Washington, DC: USCB. http://www2.census.gov/prod2/statcomp/documents/CT1970p1-01.pdf (accessed February 20, 2015).

USDA (United States Department of Agriculture). 2010. Wood Handbook. Madison: USDA. www.fpl.fs.fed.us/documnts/fplgtr/fpl_gtr190.pdf (accessed February 20, 2015).

USDOE (US Department of Energy). 2009. Modern Shale Gas Development in the United States: A Primer. Washington, DC: USDOE. http://energy.gov/sites/prod/files/2013/03/f0/ShaleGasPrimer_Online_4-2009.pdf (accessed February 20, 2015).

USEIA (US Energy Information Administration). 2008. About U.S. natural gas pipelines. http://www.eia.gov/pub/oil_gas/natural_gas/analysis_publications/ngpipeline/index.html (accessed February 20, 2015).

USEIA. 2010. Major tight gas plays, lower 48 states. http://www.eia.gov/oil_gas/rpd/tight_gas.pdf (accessed February 20, 2015).

USEIA. 2011. Over one-third of natural gas produced in North Dakota is flared or otherwise not marketed. http://www.eia.gov/todayinenergy/detail.cfm?id=4030 (accessed March 10, 2015).

USEIA. 2013a. Electric power annual. http://www.eia.gov/electricity/annual/ (accessed February 20, 2015).

USEIA. 2013b. Oil and gas industry employment growing much faster than total private sector employment. http://www.eia.gov/todayinenergy/detail.cfm?id=12451 (accessed February 20, 2015).

USEIA. 2014a. Cost of crude oil and natural gas well drilling. http://www.eia.gov/dnav/pet/pet_crd_wellcost_s1_a.htm (accessed February 20, 2015).

USEIA. 2014b. Number of gas and gas condensate wells. http://www.eia.gov/dnav/ng/hist/na1170_nus_8a.htm (accessed February 20, 2015).

USEIA. 2014c. Natural gas prices. http://www.eia.gov/naturalgas/data.cfm#prices (accessed February 20, 2015).

USEIA. 2014d. Natural gas storage. http://www.eia.gov/naturalgas/data.cfm (accessed February 20, 2015).

USEIA. 2014e. Natural gas consumption by end use. http://www.eia.gov/dnav/ng/ng_cons_sum_dcu_nus_a.htm (accessed February 20, 2015).

USEIA. 2014f. Natural gas-fired combustion turbines are generally used to meet peak electricity load. http://www.eia.gov/todayinenergy/detail.cfm?id=13191 (accessed February 20, 2015).

USEIA. 2014g. Levelized cost and levelized avoided cost of new generation resources in the Annual Energy Outlook 2014. http://www.eia.gov/forecasts/aeo/electricity_generation.cfm (accessed February 20, 2015).

USEIA. 2014h. U.S. natural gas imports by country. http://www.eia.gov/dnav/ng/ng_move_impc_s1_a.htm (accessed February 20, 2015).

USEIA. 2014i. Annual energy outlook. http://www.eia.gov/forecasts/aeo/data.cfm (accessed February 20, 2015).

USEIA. 2014j. Growth in U.S. hydrocarbon production from shale resources driven by drilling efficiency. http://www.eia.gov/todayinenergy/detail.cfm?id=15351 (accessed February 20, 2015).

USEIA. 2014k. Coalbed methane production. http://www.eia.gov/dnav/ng/ng_prod_coalbed_s1_a.htm (accessed February 20, 2015).

USEIA. 2014l. Bakken. http://www.eia.gov/todayinenergy/index.cfm?tg=bakken (accessed February 20, 2015).

USEIA. 2014m. Gross gas withdrawals and production. http://www.eia.gov/dnav/ng/ng_prod_sum_dcu_NUS_m.htm (accessed February 20, 2015).

USEIA. 1993. Drilling Sideways—A Review of Horizontal Drilling Technology and Its Domestic Application. Washington, DC: USEIA.

USEPA. 2008. Direct emissions from stationary combustion sources. http://www.epa.gov/climateleadership/documents/resources/stationarycombustionguidance.pdf (accessed February 20, 2015).

USEPA. 2012. Global anthropogenic non-CO2 greenhouse gas emissions: 1990–2030. http://www.epa.gov/climatechange/Downloads/EPAactivities/EPA_Global_NonCO2_Projections_Dec2012.pdf (accessed February 20, 2015).

USEPA. 2014. Inventory of U.S. Greenhouse Gas Emissions and Sinks: 1990–2012. Washington, DC: USEPA. http://www.epa.gov/climatechange/ghgemissions/usinventoryreport.html (accessed February 20, 2015).

USFPC (US Federal Power Commission). 1965. Northeast Power Failure: November 9 and 10, 1965. Washington, DC: US FPC.

USGS (United States Geological Survey). 2000. USGS world petroleum assessment 2000. http://pubs.usgs.gov/fs/fs-062-03/FS-062-03.pdf (accessed February 20, 2015).

USGS. 2011. USGS releases new assessment of gas resources in the Marcellus Shale, Appalachian Basin. http://www.usgs.gov/newsroom/article.asp?ID=2893&from=rss_home#.U9VnX7kg-M8 (accessed February 20, 2015).

USGS. 2013. Water resources and shale gas/oil production in the Appalachian Basin—Critical issues and evolving developments. http://www.usgs.gov/newsroom/article.asp?ID=2893&from=rss_home#.U9K8xbkg-M8 (accessed February 20, 2015).

USGS. 2014. Mineral commodity summaries 2014. http://minerals.usgs.gov/minerals/pubs/mcs/2014/mcs2014.pdf (accessed February 20, 2015).

USNRC (US Nuclear Regulatory Commission). 2014. Combined license applications for new reactors. http://www.nrc.gov/reactors/new-reactors/col.html (accessed February 20, 2015).

Valenti, M. 1991. Combined-cycle plants: Burning cleaner and saving fuel. Mechanical Engineering 113(9):46–50.

Van der Elst, N.J. et al. 2013. Enhanced remote earthquake triggering at fluid-injection sites in the Midwestern United States. Science 341, 164.

Vengosh, A. et al. 2013. The effects of shale gas exploration and hydraulic fracturing on the quality of water resources in the United States. Procedia Earth and Planetary Science 7:863–866.

Vidic, R. et al. 2013. Impact of shale gas development on regional water quality. Science 340:1235009.

Vincent, M.C. and M.R. Besler. 2013. Emerging best practices ensure fracture effectiveness over time in resource plays. The American Oil & Gas Reporter, December 2013:61–71.

Vogel. 1993. Great Well of China. Scientific American 268(6):116–121.

Volta, A. 1777. Lettere del Signor Don Alessandro Volta … sull'aria infiammabile nativa delle paludi. Milano: Nella stamperia di Giuseppe Marelli.

Wadham, J.L. et al. 2012. Potential methane reservoirs beneath Antarctica. Nature 488:633–637.

Wageningen UR. 2014. Wageningen UR Glastuinbouw. http://www.wageningenur.nl/nl/Expertises-Dienstverlening/Onderzoeksinstituten/wageningen-ur-glastuinbouw.htm (accessed February 20, 2015).

Wang, Z. and A. Krupnick. 2013. A Retrospective Review of Shale Gas Development in the United State: What Led to the Boom? Washington, DC: Resources for the Future. http://www.rff.org/RFF/documents/RFF-DP-13-12.pdf (accessed February 20, 2015).

WCA (World Coal Association). 2014. Coal bed methane. http://www.worldcoal.org/coal/coal-seam-methane/coal-bed-methane/ (accessed February 20, 2015).

Webber, W.H.Y. 1918. Gas & Gas Making: Growth, Methods and Prospects of the Gas Industry. Common Commodities and Industries. London: Sir Isaac Pitman & Sons.

Weil, B.H. 1949. The Technology of Fischer-Tropsch Process. London: Constable.

Weissman, A.D. 2013. U.S. natural gas industry positioned for dominant role in global LNG markets. *The American Oil & Gas Reporter*, October 2013:41–53.

Werpy, M.R. et al. 2010. Natural Gas Vehicles: Status, Barriers, and Opportunities. Argonne: Argonne National Laboratory. http://www.afdc.energy.gov/pdfs/anl_esd_10-4.pdf (accessed February 20, 2015).

Wilson, R. 1973. Natural gas is a beautiful thing? Bulletin of the Atomic Scientists 29(7):35–40.

Worrell, E. et al. 2008. World Best Practice Energy Intensity Values for Selected Industrial Sectors. Berkeley: Ernest Orlando Lawrence Berkeley National Laboratory.

WPC (World Power Conference). 1934. Statistical Yearbook of the WPC. London: WPC.

Wright, S. 2012. An unconventional bonanza. *The Economist*, July 4, 2012. http://www.economist.com/node/21558432 (accessed February 20, 2015).

Xinhua. 2013. China's coal consumption to hit 4.8 bln tons by 2020: forecast. http://news.xinhuanet.com/english/china/2013-11/24/c_132914191.htm (accessed February 20, 2015).

Yamamoto, K. and S. Dallimore. 2008. Aurora-JOGMEC-NRCan Mallik 2006-2008 Gas Hydrate Research Project progress. Fire in the Ice, 8:1–5. http://www.netl.doe.gov/File%20Library/Research/Oil-Gas/methane%20hydrates/HMNewsSummer08.pdf (accessed February 20, 2015).

Yang, H. et al. 2008. Sulige field in the Ordos Basin: Geological setting, field discovery and tight gas reservoirs. Marine and Petroleum Geology 25:387–400.

Youngquist, W. and R.C. Duncan. 2003. North American gas: Data show supply problems. Natural Resources Research 12:229–240.

Yvon-Durocher, G. et al. 2014. Methane fluxes show consistent temperature dependence across microbial to ecosystem scales. Nature 507:488–491.

Zuckerman, G. 2013. The Frackers: The Outrageous Inside Story of the New Billionaire Wildcatters. New York: Portfolio.

# Index

*Natural Gas: Fuel for the 21st Century*, First Edition. Vaclav Smil.
© 2015 John Wiley & Sons, Ltd. Published 2015 by John Wiley & Sons, Ltd.

Printed and bound by CPI Group (UK) Ltd, Croydon, CR0 4YY

27/10/2024

14580210-0003